QP 303 VIN

Structural

;u),
on,
excite
in
general
you
won't
have
a good
time

Structural Biomaterials

Revised edition

JULIAN VINCENT

PRINCETON UNIVERSITY PRESS
PRINCETON, NEW JERSEY

Published by Princeton University Press, 41 William Street, Princeton, New Jersey 08540

Macmillan edition of 1982 © Julian F. V. Vincent 1982

Library of Congress Cataloging in Publication Data

Vincent, Julian F. V.
    Structural biomaterials / by Julian Vincent. – Rev. ed.
      p. cm.
    Includes bibliographical references.
    Includes index.
    ISBN 0-691-08558-7 : – ISBN 0-691-02513-4 (pbk.) :
    1. Biopolymers – Mechanical properties.   2. Biomedical materials –
Mechanical properties.   I. Title.
    QP801.B69V56   1991
610′.28 – dc20   90-36732
                                                                CIP

Editorial and production services by Fisher Duncan,
10 Barley Mow Passage, London W4 4PH, UK

Printed in the United Kingdom

Contents

To my little daughter, Helen,
who couldn't stop me writing it

Preface

Mindful of Mark Twain's comments on how much his father learned while his son was growing up, I am cautious about ascribing all the changes I have made in this book to the results of the last ten years of research. The list of references (included by popular demand) certainly includes many books and papers from the 1980s. But there are quite a few from an earlier date whose results I did not include previously because I did not appreciate their relevance. It seems to me that the greatest advances have been made in the general area of biological ceramics. The improved understanding of both the chemistry and the ultrastructure of these materials provides much direction for mechanical analysis. The advances made in understanding the chemistry of non-ceramic biological materials seem not to be so important in furthering the understanding of their mechanical properties, although the theory of rubber elasticity seems to be in decline as a useful analogy.

But there have been some major advances. There is a general move towards the production of biomimetic and 'intelligent' materials. This involves developing new ideas for synthesis of materials to take control of morphology and chemistry closer to the molecular level at which the cell works, and understanding more about the mechanisms of control. This understanding will inevitably feed back to biology. New models, such as cellular materials, are being developed for complex materials (or are they structures?). This leads to improvements in the understanding of the mechanics of plants. The principles of biological design, in terms of shape and structure, are being applied to architecture. Biomechanics is being used increasingly by palaeontologists to expand their understanding of the lifestyle of extinct animals and plants, finding by calculation the physical limits to performance implied by a particular morphology, and hence its behavioural possibilities.

This wide-ranging use of the study of engineering design of biological materials and structures is a product of the many disciplines which that study requires. It usually happens that no single person is able to provide the knowledge and intuition for the appreciation of all the aspects of a particular problem. In the Biomechanics Group in Reading we have engineering, materials science, polymer physics, mathematics, zoology, physical chemistry, ecology, botany and agriculture more or less equally represented and of equal importance. Such a diversity of approach is stimulating to all those involved. And just as it cuts across wide boundaries of science, so it cuts across phylogenetic boundaries within biology. The biomechanics approach is largely one of systems and mechanisms. Thus the ideas involved in studying fracture or composite materials or plasticisers or ceramics can be applied equally to animals and plants, since all organisms are limited to a relatively small range of materials and structures. They are also subject to similar ranges of forces within the environment, dependent on such generalities as size. Different phyla will be found to have come to terms with these forces in different ways but, since they are all subject to the same mechanical and engineering limitations (which is why pigs can't fly . . .) the general design principles remain the same. Thus biomechanics combines with comparative functional morphology and becomes the refinement and redefinition of one of the most traditional approaches to biology. Biomechanics is neo-morphology. It is morphology plus numbers and so is 'hard' science. And in the stressful environment, mechanical functions and properties are as important as the developmental and physiological functions which generate the structures and materials.

In this book I have combined several approaches, but tried throughout to maintain a viewpoint which the biologist (and I am first and foremost a lover of animals and plants) can appreciate and understand. It seems to be necessary to provide a framework of theory, so the first chapter explains and expounds some of the basic concepts, but is as much an appendix as an introduction. There follow five chapters in which the molecular rather than the mathematical approach is developed, from proteins and polysaccharides through to ceramics. One of the outcomes of this approach is explored in a chapter on biomimetic materials, then finally there are some suggestions for you to explore the world of biomechanics for yourself. Throughout I have tried to emphasize the experimental approach and to keep all ideas within some sort of biological context. I have also, unashamedly, included only those ideas and topics which I find interesting and feel I can understand. I apologize profoundly to those whose work I appear to find, by implication of omission, boring or obscure! So this book is a statement, as well as a product, of my enthusiasm. It is therefore bound to be bitty and unbalanced but with odd thoughts and asides which I hope will stimulate and perhaps even amuse.

My thanks are due to many friends and colleagues, some of them past students. I was going to list them, but their names all appear somewhere in this volume. Even so, I must particularly mention Jim Gordon who was the first person who made this all seem possible for me.

Julian Vincent
Department of Pure and Applied Zoology
The University, Reading, UK

Technical notes: The text was typed on a Sinclair QL running text87 written by Fred Toussi of Software87. The illustrations were produced on an Atari ST running Hyperpaint written by Dimitri Koveos and printed on a Hewlett-Packard DeskJet Plus.

List of Symbols

A_o	initial cross-sectional area
A_s	area of shearing
b	$1/r$
C_1, C_2	Mooney constants
c	(as subscript) composite, Cauchy
d	diameter
E	stiffness, Young's modulus
e	strain
f	force
f	(as subscript) fibre
G	shear modulus;
G	giga (10^9)
G^*, G', G''	complex, storage and loss (shear) moduli
$G(t)$	relaxation (shear) modulus
g	acceleration due to gravity
$H(\tau)$	relaxation spectrum function
Hz	Hertz, cycles per second
H	(as subscript) Hencky
I	second moment of area
J	compliance
J	joule
$J(t)$	creep compliance
k	Boltzmann's constant, radius of gyration ($= I/A_o^2$)
k	kilo (10^3)
L_o	initial length
dl, l	change in length
M_c	molecular weight between crosslinks
M	mega (10^6)
m	mass
m	metre, milli (10^{-3})
m	(as subscript) matrix, Maxwell
N	newton

n	nano (10^{-9})
Pa	pascal (Nm^{-2})
R	resilience; the gas constant
r	end-to-end distance of a randomly orientated chain
S	entropy
T	temperature
t	time
U	internal energy
V	volume fraction
W_f	work of fracture; 'toughness'
$x'y'z'$	(as subscripts) three orthogonal axes
γ	shear strain
Δ	small change, decrement
δ	phase angle
η	viscosity
θ, ϕ, ψ	angle
λ	extension ratio ($=e+1$)
μ	micro (10^{-6})
ν	Poisson's ratio
ρ	density
σ	stress, strength
τ	shear stress, relaxation time
ω	frequency

Chapter 1

Basic Theory of Elasticity and Viscoelasticity

In the physically stressful environment there are three ways in which a material can respond to external forces. It can add the load directly onto the forces which hold the constituent atoms or molecules together. This occurs in simple crystalline (including polymeric crystalline) and ceramic materials; such materials are typically very rigid. Or it can feed the energy into large changes in shape (the main mechanism in rubber and other non-crystalline polymers). And finally it can flow away from the force and deform either semi-permanently (as with viscoelastic materials) or permanently (as with plastic materials).

1.1 HOOKEAN MATERIALS AND SHORT-RANGE FORCES

The first class of materials is exemplified amongst biological materials by bone and shell (Chapter 6), by the cellulose of plant cell walls (Chapter 3), by the cell walls of diatoms, by the crystalline parts of a silk thread (Chapter 2), and by the chitin of arthropod skeletons (Chapter 5). All of these materials have a well-ordered and tightly bonded structure and broadly fall into the same class of material as metals and glasses. What happens when such materials are loaded, as when a muscle pulls on a bone, or when a shark crunches its way through its victim's leg?

A suitable engineering text for the following is Richards (1961).

In a material at equilibrium, in the unloaded state, the distance between adjacent atoms is 0.1 to 0.2 nm. At this interatomic distance the forces of repulsion between two adjacent atoms balance the forces of attraction. When the material is stretched or compressed the atoms are forced out of their equilibrium positions and are either parted or are brought together until the forces generated between them, either of attraction or repulsion respectively, balance the external force (Fig. 1.1). Note that the line is nearly straight for a fair distance on either side of the origin and that it curves eventually on the compression side (the repulsion forces obey an inverse square law) and on the

1

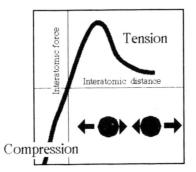

Figure 1.1 Stress–strain curve at the atomic level for a rigid material- the idealized curve for a 'perfect' material. The origin represents the equilibrium interatomic distance. On either side of the origin the curve is nearly straight.

extension side. With most stiff materials the extension or compression is limited by other factors (see Section 1.6) to less than 1% of the bond length, frequently less, so that the relationship between force and distance is, to all intents and purposes, linear. When the load is removed, the interatomic forces take the atoms back to their original equilibrium positions.

It is a fairly simple exercise to extend this relationship to a material such as a crystal of hydroxyapatite in a bone. This crystal consists of a large number of atoms held together by bonds. The behaviour of the entire crystal in response to the force is the summed responses of the individual bonds. Thus one arrives at the phenomenon known as Hooke's law, originally stated (anagrammatically) as *ut tensio, sic vis* – as the extension, so the force. In other words extension and force are directly and simply proportional to one another and this is a direct outcome of the behaviour of the interatomic bond. But when dealing with a piece of material it is obvious that measurements cannot conveniently be made of the interatomic distance (though they have been made using X-ray diffraction, which confirms the following). What is actually measured is the increase in length of the whole sample or a part of the sample (making the verifiable assumption that in a homogeneous material one part will deform as much as the next). This is then expressed as a function of the starting length called the *strain* (epsilon).

Strain can be expressed in a number of ways, each offering certain advantages and insights into the processes of deformation. The most commonly encountered form is *conventional, nominal, engineering* or Cauchy strain, which is the increase in length per unit starting length.

$$e_C = \frac{\Delta l}{L_0} \qquad \text{[Eq.1.1]}$$

This estimate of extension works well if the material is extended by no more than a tenth of its starting length. Strain is expressed either (as in this book) as a number (e.g. 0.005) or as a percentage (e.g. 0.5%).

The force acting on each bond is a function of the number of bonds available to share the load. So if the area over which the force acts is doubled, then the load carried by each bond will be halved. Thus it is important, if one is to bring the data to the (notionally) irreducible level of the atomic bond, to express the force as a function of the number of bonds which are responding to it. In practice this means expressing the force as force per unit area, which is called the *stress* (σ: sigma).

$$\sigma = \frac{f}{A_0}$$ [Eq.1.2]

However, just as with strain, this simple equation is suitable only for small extensions.

In SI units, the force is expressed in Newtons (a function of mass and the acceleration due to gravity: one Newton is approximately the force due to 100 g, the weight of an average apple), the area in square metres. One Newton acting over an area of one square metre is a Pascal (Pa). Other units are in use in many parts of the world. For instance, in the USA the unit of force is the dyne (the force exerted by one gram under the influence of gravity) and the unit of area is the square centimetre. One dyne per square centimetre is one hundred thousandth of a Pascal. Traditional engineers in Britain often use pounds and square inches as their measures of 'force' and area.

The slope of the straight, or Hookean, part of the curve in Fig. 1.1 is characteristic of the bond type and is a function of the energy of the bond. For the same reason it is found that the ratio of stress to strain is more or less characteristic of a material. This ratio is the stiffness or Young's modulus (E):

$$E = \frac{\sigma}{e}$$ [Eq.1.3]

The units of E are the same as for stress, since strain is a pure number. Graphs showing the relationship between stress and strain are conveniently plotted with the strain axis horizontal and the stress axis vertical, irrespective of whether the relationship was determined by stretching the test piece in a machine and recording the developed forces or by hanging masses onto the test piece and recording the extension. Do not be surprised if it takes a long time for the mental distinctions between stress and strain to become totally clear. Not only are the concepts surprisingly difficult to disentangle, but the confusion is compounded by the uncritical usage of stress and strain in everyday speech.

One other characteristic of Hookean materials is that they are elastic. That is to say, they can be deformed (within limits) and will return to their original shape almost immediately the force is removed (almost immediately because the stress wave travels through the material at the speed of sound in that material. So when you pull on the brake lever on your bicycle the brake blocks begin

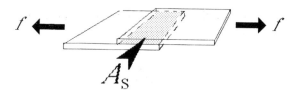

Figure 1.2 Conditions for the definition of shear stress (Eq.1.4).

to move a short time later, the time dependent partly on the speed of sound in the steel cable and partly on the length of the cable). This use of the word 'elastic' must not be confused with the usage of the term as in 'elastic band' where elastic is taken to mean highly extensible.

Young's modulus is a measure of stiffness in simple extension or compression. There are ways of deforming a material which have different effects on the interatomic forces and therefore different effects on the material. Such a mode of deformation, frequently met, is shear. (Another mode of deformation – volume change from which is derived the bulk modulus – is ignored here.) As with Young's modulus, the shear modulus is defined as the ratio of stress to strain. The shear stress (τ: tau) is defined as (see Fig. 1.2):

$$\tau = \frac{f}{A_S}$$ [Eq.1.4]

The shear strain is defined somewhat differently (Fig. 1.3). The strain, y, is measured in radians and the shear modulus (G) is given by:

$$G = \frac{\tau}{y}$$ [Eq.1.5]

The simple picture given here is for isotropic materials whose structure and, therefore, mechanical response are the same in all directions. The Young's

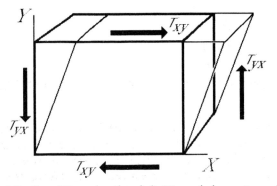

Figure 1.3 Conditions for the definition of shear stress (Eq.1.5).

modulus and the shear modulus in an isotropic material can be related to each other by the expression:

$$G = \frac{E}{2(1 + \nu)}$$ [Eq.1.6]

where ν is Poisson's ratio. This important ratio is discussed at greater length in Section 4.3. A material which is Hookean in extension is usually Hookean in shear. The mathematics for high strain shear deformation are not considered here, and indeed remain to be established!

1.2   NON-HOOKEAN MATERIALS—HIGH STRAINS

With greater deformation, another form of strain – true or Hencky strain – is a better indicator of what is going on in the material. With true strain, each small extension is expressed as a fraction of the immediately preceding or instantaneous length. It is slightly more cumbersome to calculate

$$l_H = \ln(l/l) = \ln(l + l_c)$$ [Eq.1.1a]

and has the curious property that the sample does not 'remember' its strain history. It is an instantaneous measure of strain. Figure 1.4 compares true and conventional strain, showing that the mutual deviation is far greater in compression.

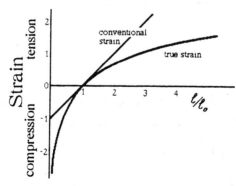

Figure 1.4 Comparison of true and conventional strain plotted against extension ratio.

At larger extensions (greater than 0.1 or so), Poisson's ratio effects (see Section 4.3) will cause the sample to become narrower, reducing the area over which the force is being transmitted. This will cause the true stress to increase at a higher rate than the conventional stress (Fig. 1.5). However since, as will be seen, Poisson's ratio frequently varies with strain, especially with soft biological

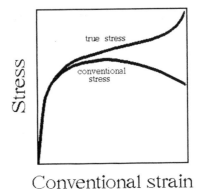

Conventional strain

Figure 1.5 Comparison of true and conventional stress plotted against conventional strain.

materials which are complex, extensible and fibrous, it is not possible to give a universal formula for calculating true stress from the starting conditions. The cross-sectional area has to be measured at the particular strain for which the stress is to be calculated. However, just to give you the feel of the relationship between engineering stress and true stress, assume the Poisson's ratio varies in the same way as a rubber and that the volume of the material remains constant (for many biological materials a doubtful assumption – Section 4.4.2). Then if the cross-sectional area at any time is A, and A_0 the area at zero strain (L_0), then:

$$A(L_0 + \Delta l) = A_0 L_0 \qquad\qquad [\text{Eq.1.2a}]$$

so

$$A = \frac{A_0 L_0}{(L_0 + \Delta l)} = \frac{A_0}{(l + e_C)}$$

so that

$$\sigma_H = \frac{f}{A} = \frac{(l + e_C)f}{A_0} = (l + e_C)\sigma_C$$

which relates the true stress to the apparent or engineering stress.

Both true and conventional methods of expressing stress and strain are used; in the small-strain elastic range, the conventional measures are more usually used and are more convenient, though not strictly accurate. However, at the strains which soft biological materials reach, true stress and strain are the proper indicators of what is happening in the material, although these parameters are seldom used. When the material starts to yield (Section 1.6), even true stress and strain are inadequate, since neither is uniform across the yield zone. Even so, they are to be preferred to the conventional parameters. At present, the use of true stress and strain is a counsel of perfection. Since all biological materials show some form of relaxation (see Section 1.4), then an estimate of cross-sectional area for the calculation of true stress has to be made instantaneously. In practice, where such data are required, it is often found that the best technique is to record the test with a number of cameras using split-screen video and to make the necessary measurements of the specimen after the test is completed. This sort of practical complexity goes a long way to explaining why there are so few data on biological materials in which true stress has been measured.

1.2.1 Rubbers and long-range forces

Very few rubbers exist in nature in a pure and simple form. Obviously there is the latex of *Hevea brasilliensis*, but this is neither cross-linked (as is required if the rubber is to bear loads) nor are its rubbery properties used within the plant. There are two animal rubbers which have been investigated – resilin (found in wing-hinges, hair bases and so on in insects) and abductin (found in the inner hinge ligament of bivalve molluscs). Both these have the properties which one would expect of a rubber.

The important point about a rubber is that it is a long-chain polymer of random conformation. That is to say it is composed of very long chains (molecular weight of about 10^5) of one or more monomer units, and that each unit is more or less freely jointed into the chain so that each joint allows a wide range of movement. This is called 'free rotation' about the bonds of the backbone and is what distinguishes a rubbery polymer from a crystalline one: in a crystalline polymer (or in areas of crystallinity) the units cannot move freely because they are packed so closely, and rubbery behaviour is impossible. In actual fact it takes more than one monomer unit or residue to make a freely rotating unit or 'random link'. This is because the monomer units are of a finite size and shape and so cannot move with absolute freedom without hitting their neighbours ('steric hindrance'). With paraffin chains with a tetrahedral valence angle it takes three C-C links to make up a freely rotating or equivalent random link; with *cis*-polyisoprene units, as in *Hevea* rubber, the number of monomer units per random link is 0.77, since there are four bonds to each isoprene unit (Treloar 1975). With proteins the equivalent random link is 4 to 6 amino acids (Andersen & Weis-Fogh 1964). Under the influence of Brownian motion the free rotation of the equivalent random links about the backbone of the polymer allows the chain to assume a random conformation. In other words there is no pattern to the angles which each link makes with its neighbour other than a statistical one. The fact that the molecules are in Brownian motion also leads to the concept of kinetic freedom, which is a way of saying that the chains are free to thresh around in any direction. Brownian motion is temperature dependent – as the temperature increases so the movement of the molecules and their subunits becomes more and more frenetic. In a similar manner, reduction in temperature causes the activity to be reduced until finally, at a temperature dependent on the particular rubber in question, it ceases altogether and any force which is exerted on the rubber meets the resistance of the covalent bonds linking the atoms, probably bending rather than stretching them. A rubber at the temperature of liquid nitrogen has similar properties to the ceramic phase of bone, although it has a lower modulus, is Hookean, and is said to be glassy. The temperature at which this occurs is called the glass transition temperature (T_g).

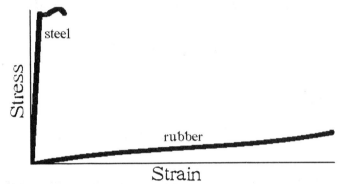

Figure 1.6 Comparison of the stress–strain curves of steel and rubber (the slopes have not been accurately calculated but give some idea of the differences - probably greater than shown here).

However, the response of rubber to stress at normal temperatures is very different (Fig. 1.6 compares the shape of curve for a rubber and a mild steel). Rather than going deeply into the thermodynamics (see Ramsay 1971 for an introduction to thermodynamics written by, and understandable by, a biologist), suffice it to say that it is possible to divide the internal reaction of a material to an external force into two components: one is due to the bonds holding the components together (as in metals and glasses) and is loosely called the internal energy term. The other is due to the mobility of the components and is called the entropy term. Entropy is a measure of the disorder of a system (life itself has been characterized as a continual battle against entropy) and the significance of this term can be appreciated as follows. When a piece of rubber is stretched the chains are pulled out of their random configuration and come to be arranged rather more parallel to the direction in which they are being extended. This partial orientation represents an increase in order and thus a decrease in entropy. Now consider the system as a kinetic one, with the rubber chains writhing in Brownian motion. It is this writhing which produces the tension. Imagine that you hold one of these writhing molecules by the ends and try to pull it straight. You are trying, by doing work on the molecule, to decrease its entropy. If the temperature increases and the molecule writhes more violently it opposes your efforts with greater force.

These ideas can be enshrined in the following mathematical expressions whose derivations are not considered here but which are described in Wainwright *et al.* (1976) and Treloar (1975). First the thermodynamic equation of state:

$$f = \left(\frac{dU}{dl}\right)_T - T\left(\frac{dS}{dl}\right)_T \qquad \text{[Eq.1.7]}$$

This states that the force is distributed between a change in internal energy (dU/dl) and a (temperature-dependent) change in entropy (dS/dl). In addition it can be shown that:

$$\left(\frac{dS}{dl}\right)_T = -\left(\frac{df}{dT}\right)_l \qquad \text{[Eq.1.8]}$$

and therefore that

$$\left(\frac{dU}{dl}\right)_T = f - T\left(\frac{df}{dT}\right)_l \qquad \text{[Eq.1.9]}$$

Equations 1.7 and 1.9 are of fundamental importance in describing rubber elasticity since they provide a direct means of experimental determination of the contributions to the restoring force generated by the changes in internal energy (U) and entropy (S) on deformation. All that is needed is a set of equilibrium values for the tension at constant length over the range of temperatures. This has been done for resilin by Weis-Fogh (1961a,b) (Fig. 1.7, 1.8). It has also been done for abductin by Alexander (1966) although his published data are not so complete and are omitted here. The importance of these experiments is that they provide strong evidence for the rubbery nature of resilin and abductin. Obviously the above experiment has also been performed on the

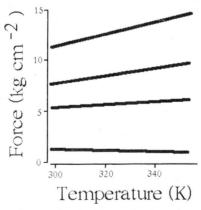

Figure 1.7 Results from thermodynamic experiments on resilin showing the force required to maintain constant extension at different extension ratios (Weis-Fogh 1961).

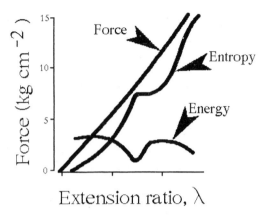

Figure 1.8 Results from thermodynamic experiments on resilin showing the contributions of entropy and internal energy to the elastic behaviour (at 300 K) (Weis-Fogh 1961).

cross-linked (vulcanized) rubber of *Hevea* but not on the latex since this is not cross-linked so it is impossible to obtain equilibrium values of the load. Such cross-links may be of various kinds, both covalent (S-S in vulcanized rubber and keratin, phenolics in resilin and possibly other invertebrate protein matrices) and non-covalent (most commonly H-bonds). The covalent cross-links have high energy and can support stresses to the same degree as the bonds in the polymer backbone. The degree of cross-linking affects the properties of the material by controlling the degree to which the material can reorganize itself in response to stress. If there is too little cross-linking there will be permanent 'set' after the load is removed; if too much there will be no rubber elasticity. The degree of cross-linking in soft rubbers is about 1% implying that there are about 100 chain links between ties.

There are other useful relationships which have been worked out for rubbers, each using a different starting point. They are none of them particularly satisfactory, especially at high strains where there is evidence that the rubber is crystallizing and so effectively becoming a different material. Using as a model single flexible random coil chain, freely moving and randomly orientated in space, the following relationship can be derived.

$$f = 2kTb^2r \qquad\qquad [\text{Eq.1.10}]$$

In the derivation of this equation, it is found that b is inversely proportional to r. So the force is inversely proportional to the end-to-end distance of the chain, and the shorter the chain the stiffer the material. Another useful formula derived from a statistical treatment of rubber states that:

$$f = \rho RT/M_c(\lambda - \lambda^{-2}) \qquad\qquad [\text{Eq.1.11}]$$

$$\rho RT/M_c = G \qquad\qquad [\text{Eq.1.11a}]$$

This again shows that the stiffness of the material is inversely proportional to the free length of the molecular chains. The trouble with these relationships is that they are applicable only for small strains of 0.3 or so. At higher extensions, especially in *Hevea* rubbers, orientation effects and strain crystallization (see below) of the polymer chains cause deviations from the theoretical force and the actual forces are higher than predicted.

Another approach to rubber elasticity which has been useful is purely phenomenological. That is to say the material is treated a a black box, some assumptions made about it, and no knowledge required of the molecular responses in the black box. Mooney has derived the following relationship for rubber in simple tension:

$$f = 2(\lambda - \lambda^{-2})(C_1 + C_2/\lambda) \qquad \text{[Eq.1.12]}$$

This is related to Eq.1.11 by Eq.1.13:

$$G = 2(C_2 + C_2) = \rho RT/M_c \qquad \text{[Eq.1.13]}$$

and the statistical theory represents the special case of the Mooney theory for which $C_2 = 0$. Experiment suggests that C_1 is related to the degree of cross-linking and that C_2 is related to the degree of swelling. Since, as we shall see, the swelling volume of biological elastomers can vary both during development and among comparable materials in use, the Mooney relationship could offer some interesting insights into the development and working of biomaterials.

Ultimately, all this talk of rubber elasticity may be of little relevance to biological materials since so few biological materials have been shown to be rubbery. There are one or two horror stories in which materials have conformed to some or all of the above criteria of rubber elasticity but have subsequently been shown to rely on different, usually more specific, molecular mechanisms. This problem is discussed in more detail at the end of the next chapter.

1.3   VISCOELASTICITY – STRESS, STRAIN AND TIME

There are many biological materials which contain crystalline components; there are a few which contain rubbers which are sufficiently well cross-linked to be analysed in terms of rubber elasticity. But by far the greatest number, if not at all, biological materials are viscoelastic to a greater or lesser extent. They have a viscous component. The mathematical description of such materials involves the introduction of a new variable – time – so that while crystalline materials and 'ideal' rubbers, at constant temperature, have their mechanical properties described in terms of stress and strain, viscoelastic materials need the additional term of time in their description. Viscoelasticity and related phenomena are of great importance in the study of biological materials. Just

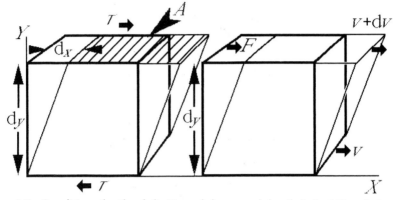

Figure 1.9 Conditions for the definition of shear modulus (left, Eq.1.5) and viscosity (right, Eq.1.4).

as strain can be measured in more than one way, so the related rate of strain (i.e. the amount of strain per unit time) can be measured in a number of different ways (Whorlow 1980). Cauchy strain rate is given by $dl/l_0 dt$; Hencky strain rate by $dl/l dt$. In each expression, dl is the infinitesimally small extension achieved during the short time dt, l_0 is the starting length at zero time, l is the length just before the present extension.

Viscosity is defined as the ratio of shearing stress to velocity gradient (Newton's law). Its equivalence to the shear modulus can be seen in Fig. 1.9; its definition is:

$$G = \frac{\tau}{y} = \frac{F/A}{dx/dy} \ [1.5]; \ \eta = \frac{F/A}{dv/dy} \qquad [\text{Eq.1.14}]$$

'Newtonian' viscosity is independent of strain or shear rate. This means that if the force applied to a Newtonian fluid is doubled, the shear rate will also be doubled. Non-Newtonian fluids are those which respond with a more, or less, than doubled shear rate, depending on whether or not they show shear thinning or shear thickening. Most biological materials show shear thinning so that doubling the force will more than double the shear rate, thus making deformation of the material relatively easier at higher shear rates. The units of viscosity are $\text{kg m}^{-1}\text{s}^{-1}$ or N s m^{-1}. At this point it is necessary to say that viscoelasticity is not plasticity, with which it is often confused. A viscoelastic material will return to its original shape after any deforming force has been removed (i.e. show an elastic response) even though it will take time to do so (i.e. have a viscous component to the response). A plastic material will not return to its original shape after the removal of the load. In metals, plasticity is call ductility. It is, if you like, the converse of elasticity in that the energy of deformation is not stored but is entirely dissipated. A material can show a combination of elasticity and plasticity, in which case although it returns part

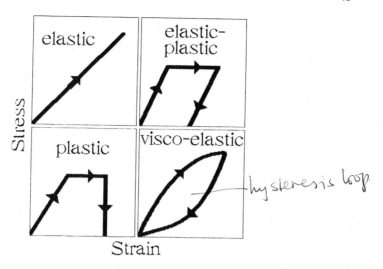

Figure 1.10 Stress-strain curves illustrating different types of behaviour.

of the way to its original shape on removal of the load, some permanent deformation or 'set' remains. This is due to plastic deformation or molecular 'slippage' of an irreversible nature (Fig. 1.10).

There are two major types of experiment performed on viscoelastic materials. *Transient* experiments involve deforming the material (by simple elongation or in shear) and following the response of the material as time goes by. There are two transient experiments. In one the material is loaded and the change of deformation with time is noted. This is the creep experiment. Under load, segments of the molecules of the material rotate and flow relative to each other

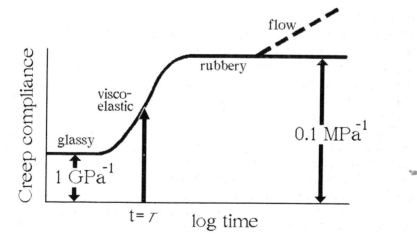

Figure 1.11 Creep compliance, *J(t)*, as a function of time, *t*. The characteristic or retardation time is *τ'*.

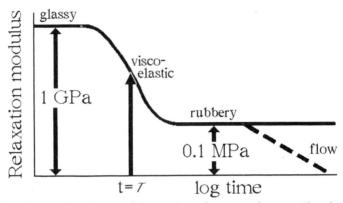

Figure 1.12 Stress–relaxation modulus $G(t)$ as a function of time, t. The characteristic or relaxation time is τ.

at a rate controlled by the viscosity of the material, the stress, the temperature and the time for which the material has been stressed. Figure 1.11 shows how the strain varies with constant (engineering) stress over a wide range of times after loading. The parameter, J, obtained by dividing the strain by the stress, is the *compliance* (roughly the inverse, or opposite, of stiffness) and is here further defined as the *creep compliance* ($J(t)$). A compliant (or pliant) material is a non-stiff or soggy material. A hi-fi pick-up cartridge with a 'high compliance' has the stylus mounted in a flexible material so that it will present little resistance to being moved by the irregularities which comprise the signal on the groove of the record. The molecular origin of the various regions of the compliance curve are discussed in Section 1.5.

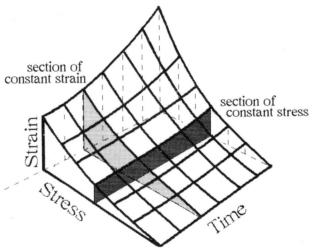

Figure 1.13 Three-dimensional representation showing the interrelationship between stress, strain and time. Derived from isochronal stress–strain curves along the stress axis and creep curves along the time axis (Ward 1971).

The other transient experiment is the *stress-relaxation* experiment in which the material is deformed and the force required to maintain the deformation at a constant value is measured as, with time, the molecules of the material move relative to each other. Thus the stress required to hold the material at constant deformation dies away with time and is said to relax. Figure 1.12 shows how the stress varies with constant (engineering) strain (the relaxation modulus, $E(t)$ for simple extension, $G(t)$ for shear) in a manner analogous to that for creep compliance. Note that the two transient experiments are possible because there are three variables – stress, strain and time. It is therefore possible to plot a three-dimensional surface showing how these variable are interrelated (Fig. 1.13 illustrating creep).

The other major type of experiment is the *dynamic* one in which either stress or strain (usually strain) is varied cyclically (usually sinusoidally for mathematical convenience) with time and the response measured at various different frequencies of deformation.

Transient experiments are usually easier to understand and will be described first. The assumptions which are made about the mechanical response of the material are similar for both transient and dynamic experiments.

There are three major ways of describing viscoelastic behaviour, all interrelated. The first starts with the Boltzmann superposition principle and is sometimes called the *integral* representation of linear viscoelasticity because it defines an integral equation. The second way, which leads to a linear differential equation and is therefore called the *differential* representation, uses assemblages of Hookean springs and Newtonian viscous elements (dashpots) as models. The third method is based on assumptions about the molecules themselves. At this point you may find it easier to read the section on the behaviour of the molecules of viscoelastic materials and then come to the phenomenological approach which follows. Either way you will need to read the following sections several times to see how all the different measurements and ideas fit together.

It cannot be emphasized too much that both the integral and differential models are *only* models and are not explanations. A number of papers on biological materials interpret the behaviour of the material solely and finally in terms of springs and dashpots, as if that were an answer. The models are like the water diviner's twig or the hangman's noose – they serve to concentrate the mind. With the use of mathematical expressions derived from consideration of the models it is possible to derive constants which can be used as a basis for comparison or prediction. But it is highly unlikely that a biological material can be described in terms of a single spring and dashpot unit. The other major caveat in all the theory of viscoelasticity that follows is that both the models and their mathematical representations rely on *linearity* of response of both elastic and viscous components. This is normally considered to be attainable only at strains less (usually much less) than 0.01. But nearly all biological materials (and most artificial polymers) are not only non-linear in response but normally function at high and extremely high (0.5 +) strains. The models for viscoelasticity are not valid under these conditions. This is a severe limitation and one that is not commonly recognized. Thus much work on artificial and natural polymers is of dubious value since it is applying linear, small strain models to non-linear, large strain materials. The fact that such data may well often be internally consistent is no argument for the acceptance of the linear interpretation; it may merely be coincidence. The mathematics of viscoelasticity at large strains remain to be worked out.

1.4 LINEAR VISCOELASTICITY

1.4.1 The integral model

The Boltzmann superposition theory may be stated as:
1. The creep in a specimen is a function of the entire loading history.
2. Each increment of load makes an independent and additive contribution to the total deformation.

(For creep, substitute stress–relaxation to cover all circumstances.) The first condition could be called the memory function: the response of the material is influenced by what has happened to it so far, so that it is 'remembering' deformations long past and allowing them to influence its present behaviour. The second condition states that if a specimen is loaded and is creeping under load, then the addition of an extra load will produce exactly the same additional creep that it would have done had that load been applied to the unloaded specimen and the specimen allowed to creep for the same amount of time. This is said to be a linear (i.e. directly additive) response. The second condition also implies that when the load is removed the recovery in length of the specimen will follow the same time course as, and be identical with, the initial creep response. The importance of Boltzmann's principle to the study of viscoelasticity is not so much that it provides any explanations as that it provides a starting point for mathematical models which can be tested against reality and refined to give a better fit. For instance many papers have been written in which the effects of different sequences of stressing or straining have been calculated according to the Boltzmann principle and the results tested by a variety of experiments on real materials. The mathematical formulation of viscoelastic behaviour derived from the Boltzmann principle is illustrated by Fig. 1.14. The total strain at time t is given by:

$$e(t) = \Delta\sigma_1 J(t - \tau_1) + \Delta\sigma_2 J(t - \tau_2) + \Delta\sigma_3 J(t - \tau_3) \qquad [\text{Eq.}1.15]$$

where J is the compliance of the material and $J(t - \tau_n)$ is the creep compliance function and is the first explicit introduction of time into these equations as an extra variable. This equation can be generalized to give:

$$e(t) = \int_{\infty}^{t} J(t - \tau_n) \, d\sigma(\tau_n) \qquad [\text{Eq.}1.15a]$$

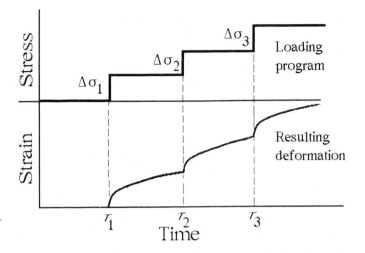

Figure 1.14 Creep behaviour of an ideal viscoelastic solid (Eq.1.5).

This is usually rewritten by removing the immediate elastic response, e, and allowing the mathematicians to rewrite the integral in what they find a more acceptable form:

$$e(t) = \left[\frac{\sigma}{G_u} \right] + \int_{\infty}^{t} J(t - \tau_n) \frac{d\sigma(\tau_n)}{d\tau_n} d\tau_n \qquad [\text{Eq.1.16}]$$

where G_u is the immediate or unrelaxed stiffness.

This divides the equation into a time-independent and a time-dependent (the integral) function. The stress-relaxation modulus can be calculated in an exactly similar manner to give:

$$\sigma(t) = [G_r e] + \int_{\infty}^{t} (t - \tau_r) \frac{de(\tau_r)}{d\tau_r} d\tau_r \qquad [\text{Eq.1.17}]$$

Notice in particular the pattern and symmetry of Eqs 1.16 and 1.17. This implies that there is probably a formal relationship between the two expressions, but not only is this relationship rather too simple and generalized to be of much use when dealing with biological materials, it is more easily approached from a different starting point!

1.4.2. The differential model

Probably the best starting point, certainly the one most easily appreciated by most biologists, is that of mechanical models using springs (elastic elements) and dashpots (viscous elements) – the differential approach. The springs are

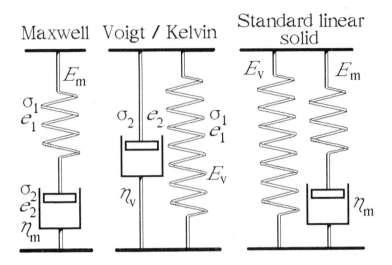

Figure 1.15 Spring and dashpot models.

Hookean and the dashpots Newtonian. The Maxwell model (Fig. 1.15) has two elements:

for the spring

$$\sigma_1 = E_m e_1 \qquad\qquad \text{[Eq.1.18a]}$$

for the dashpot

$$\sigma_2 = \eta_m \mathrm{d}e_e/\mathrm{d}t \qquad\qquad \text{[Eq.1.18b]}$$

Equation 1.18a can be divided on both sides by dt and rewritten as:

$$\frac{\mathrm{d}\sigma_1}{\mathrm{d}t} \; \frac{1}{E_m} = \frac{\mathrm{d}e_1}{\mathrm{d}t}$$

and Eq.1.17b can be rewritten to give:

$$\frac{\sigma_2}{\eta_m} = \frac{\mathrm{d}e_2}{\mathrm{d}t}$$

The two elements are in series, so $\sigma_1 = \sigma_2 = \sigma$. Also the total strain on the model, e, is the sum of e_1 and e_2. So Eqs 1.18a and 1.18b can be added to give:

$$\frac{\mathrm{d}\sigma}{\mathrm{d}t} \; \frac{1}{E_m} + \frac{\sigma}{\eta_m} = \frac{\mathrm{d}e_1}{\mathrm{d}t} + \frac{\mathrm{d}e_2}{\mathrm{d}t} = \frac{\mathrm{d}e}{\mathrm{d}t} \qquad\qquad \text{[Eq.1.19]}$$

In a stress-relaxation experiment the length is held constant, so:

$$\frac{de}{dt} = 0 \text{ and } \frac{d\sigma}{dt} \cdot \frac{1}{E_m} + \frac{\sigma}{\eta_m} = 0 \qquad [\text{Eq.1.20}]$$

so that simple rearrangement gives

$$\frac{d\sigma}{\sigma} = dt\left(\frac{-E_m}{\eta_m}\right)dt \qquad [\text{Eq.1.20a}]$$

At the start of the stress-relaxation experiment, $t = 0$ and $\sigma = \sigma_0$, the initial stress. Integrating the last equation gives:

$$\sigma = \sigma_0 \exp\left(\frac{-E_m}{\eta_m}\right) \qquad [\text{Eq.1.21}]$$

In other words the stress decays exponentially ($=$ logarithmically) with a characteristic time constant $\tau = \eta_m/E_m$ so that

$$\sigma = \sigma_0 \exp\left(\frac{-t}{\tau}\right) \qquad [\text{Eq.1.22}]$$

The Kelvin or Voigt model (Fig. 1.15) models the creep test and, using arguments similar to those with the Maxwell model, gives rise to the expression:

$$e = e_0 \exp\left(\frac{-t}{\tau}\right) \qquad [\text{Eq.1.23}]$$

Note again the extreme symmetry between the two expressions for stress-relaxation and creep. But why use two models? The Maxwell model is no use for modelling creep since under constant load the dashpot will allow viscous flow and the spring will be in constant tension. All that will then be observed is the Newtonian nature of the fluid in the dashpot. This does not accord with observation of real creep experiments so the Maxwell model is inappropriate for their description. An even more serious objection arises against the use of the Voigt model for stress-relaxation experiments, since under such conditions it behaves as an elastic solid. These objections can be overcome by combining the two models into a standard linear solid model (Fig. 1.15). However, once you start doing this sort of thing you can go on forever with more and more complex combinations of units which do not produce any more unifying concepts. Further developments along these lines are ignored here.

The most profitable approach with spring-and-dashpot models, at least when modelling artificial polymers, has been found to be that of combining numbers of Maxwell or Voigt elements (not mixing them) to obtain a spectrum of time

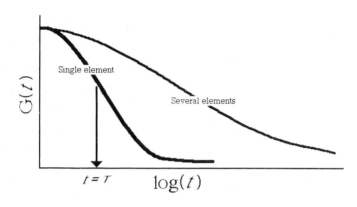

Figure 1.16 The time course of relaxation of a single Maxwell element (thick line) and of an assemblage of Maxwell elements with different relaxation times (thin line).

characteristics. If the course of relaxation of a single Maxwell element is plotted it is found to have the general shape shown in Fig. 1.16 (solid line). If the slope of this curve is plotted against log (or ln) time, a curve of shape similar to a skew log-normal distribution is obtained (Fig. 1.17, solid line). The vertical axis of Fig. 1.17 is labelled $-H(\tau)$ which is rather confusingly known as the relaxation spectrum function. The actual spectrum is the skew normal curve, and the relaxation time, τ, of the Maxwell element which generated it is given by the mode of this curve. Now if more Maxwell elements, each with a different time constant, T, are arranged in parallel, it is not difficult to see that the decay of stress will be spread over a longer period as a result of a broader spectrum of relaxation times (Figs 1.16, 1.17, broken lines). The peculiar usefulness of this relaxation spectrum is, as will be explained, that it can be derived from different types of experiment and so is a convenient transform for general comparisons between materials and tests. The other usefulness of the relaxation

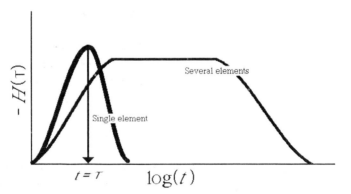

Figure 1.17 Relaxation spectrum function $[-H(\tau)]$ derived from the curves in Fig. 1.16 using the Alfrey approximation (Eq.1.29).

spectrum (and, it should be added, the related retardation spectrum calculated in a similar manner from creep data) is that it gives some idea of the number and nature of the relaxation processes going on while the stresses are relaxing. This is because each process has its own characteristic relaxation time (Section 1.5). In general, biological materials have a very broad relaxation/retardation spectrum, but Alexander (1962) has shown that the mesogloea of two sea anemones, *Calliactis parasitica* and *Metridium senile*, has a retardation spectrum dominated by a process having a retardation time of $10^{3.4}$ s (see Fig. 4.6). The model he used was a single Voigt element and he was working at strains much greater than 0.01 (3, in fact) so at least in terms of strain, the model was probably inappropriate. Bill Biggs (unpublished) reworked Alexander's data and found that they were better fitted by a five-element model with retardation times in the range of 1 to 10^4 seconds. But on plotting $-H(\tau)$ against ln t, Biggs found that the compliance is dominated by processes which have a retardation time of $10^{3.5}$ s. Although even here there is definitely more than one process involved, a single process is dominating the response. This is very unusual with biological materials and anemone mesogloea could be a good model medium for investigating the mechanisms controlling the retardation spectra of biological materials.

The relaxation spectrum is clearly an important measure of viscoelastic behaviour. Its mathematical derivation is as follows: for stress–relaxation at constant strain a single Maxwell element gives $\sigma(t) = E_m e$ exp $(-t/\tau)$. For a number of such elements joined in parallel, all at strain e, the stress is:

$$\sigma(t) = e \sum^n E_n \exp(t/\tau_n) \qquad [\text{Eq.1.24}]$$

where E_n and τ_n are the stiffness and relaxation time respectively of the nth element. Equation 1.21 can be rewritten as:

$$\sigma(t) = [G_r e] + e \int_0^\infty f(\tau) \exp(-t/\tau_n) d\tau \qquad [\text{Eq.1.25}]$$

The term $G_r e$ is the *instantaneous stress*. The integral represents the way in which the stress dies away with time, to give $\sigma(t)$. The function $f(\tau) d\tau$ replaces E_n and defines the concentration of Maxwell elements with relaxation times between τ and $(\tau + d\tau)$. The relaxation modulus is then given by:

$$G(t) = [G_r] + e \int_0^\infty f(\tau) \exp(-t/\tau) d\tau \qquad [\text{Eq.1.26}]$$

The 'relaxation time spectrum' $f(\tau)$ is replaced by $H(\tau)$ on a logarithmic time scale (simply because a log time scale is more convenient to handle). Then

$$G(t) = [G_r] + \int_0^\infty H(\tau) \exp(-t/\tau) d \ln \tau \qquad [\text{Eq.1.27}]$$

In other words the modulus at time t after the imposition of the strain is the sum of the initial modulus [initial stress divided by the (constant) strain] and of a function which describes how τ varies with time after the start of the experiment. Since τ is the ratio of the (Newtonian) viscosity to the stiffness of the individual elements, then the integral can be considered as a function of modulus with time, and describes the way in which the modulus changes (diminishes) with time. In order to calculate $H(\tau)$ simply, the Alfrey approximation is used. This assumes that $\exp(-t/\tau) = 0$ up to time $t = \tau$, and $\exp(-t/\tau) = 1$ when τ is greater than t, thus replacing a set of exponentials with a set of step functions. Equation 1.27 can then be rewritten:

$$G(t) = [G_r] + \int_0^\infty H(\tau)\mathrm{d} \ln \tau \qquad [\text{Eq.1.28}]$$

so that

$$H(\tau) = - \left[\frac{\mathrm{d}G(t)}{\mathrm{d} \ln t}\right]_{t=\tau} \qquad [\text{Eq.1.29}]$$

which is the negative slope of a plot of relaxation modulus against ln (or log) t.

1.4.3 The molecular model

The third major approach to understanding viscoelasticity is the molecular one. It is probably more convenient in this approach not to use stress–relaxation experiments or creep experiments, but to use dynamic tests rather more. This is not to say that transient experiments are of limited use. Far from it. Their versatility can be increased by varying the temperature and this will be referred to again once it has been dealt with in conjunction with dynamic tests. The strengths of stress–relaxation and creep tests are their ease of execution – measurement and experimental apparatus are very easily managed – and their immediate applicability to the life of the animal or plant. But rather more can be accomplished with dynamic testing since it is more versatile and covers a wider range of conditions. The theory is also applicable to a wide range of test rig geometries.

Once again the argument is for linear viscoelastic solids [the usual subterfuge if you have a non-linear solid (as are most biological materials) is to say that if you deform the material by a sufficiently small amount then the material will give a linear response]. Dynamic testing is particularly suitable for tests under such limitations. The sample is subjected to strain varying sinusoidally with time at a frequency ω. If the material being tested is Hookean, then the stress will be proportional to the strain – Fig. 1.18 shows stress and strain plotted against time, Fig. 1.19 (left) shows stress plotted against strain, which is a straight line. But if the material is viscous and has no elastic component, the stress in the

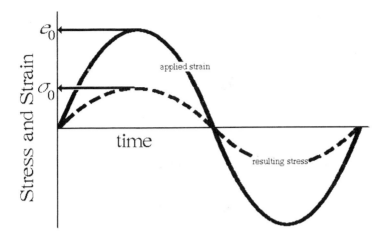

Figure 1.18 Sinusoidal strain and resulting sinusoidal stress induced in a sample undergoing dynamic testing.

material will be highest at the highest strain *rate*. Since the strain is varying sinusoidally about zero, the highest strain rate will be at zero strain. Stress in the material will be lowest at the lowest strain rate which will be the point at which the strain is highest. The resulting stress–strain Lissajous figure will be a circle. Looked at another way, the stress in a viscous material induced by sinusoidal strain is proportional to the changes of accelerations in strain, and is therefore the first differential. So if $y = \sin x$, then $\mathrm{d}y/\mathrm{d}x = \cos x$. But $\cos x$ lags 90° behind $\sin x$, so the Lissajous figure is a circle.

A viscoelastic material has a response which is partly viscous and partly elastic, so its response to a sinusoidally varying strain will be a combination of the above two extremes (Fig. 1.19, right). The problem is how to extract the information

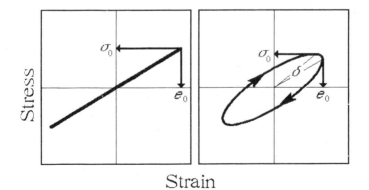

Figure 1.19 Simple vector diagrams showing (left) the data in Fig. 1.18 and (right) a viscoelastic material undergoing dynamic testing.

from the Lissajous figures. Essentially it is possible to measure a 'modulus' at the highest strain and the highest strain rate – these moduli are the elastic (or real, or storage) modulus and the viscous (or imaginary, or loss) modulus. Since the modulus, G^*, is the ratio of maximum stress to maximum strain, then

$$\sigma_0 = e_0 G^* \sin(\omega t + \delta) \qquad \text{[Eq.1.30]}$$

This expression can be expanded to give:

$$\sigma_0 = e_0(G^* \cos\delta) \sin \omega t + e_0(G^* \sin\delta) \cos \omega t$$

$$= e_0 G' \sin \omega t \qquad + e_0 G'' \cos \omega t \qquad \text{[Eq.1.31]}$$

where G' is the elastic modulus and G'' is the viscous modulus. There is thus a simple relationship between G^*, G', G'' and δ which is summarized as a vector diagram (Fig. 1.20). An exactly similar argument can be used to extract a complex compliance, J^*, and to resolve it into its components.

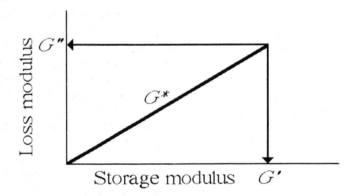

Figure 1.20 Geometrical resolution of the complex modulus, G^*, into its components, G', G'' and tan δ.

The viscosity, η, can, as would be expected, be extracted from the above analysis very simply:

$$\eta' = G''/\omega, \ \eta'' = G'/\omega \qquad \text{[Eq.1.32]}$$

The in-phase or real component of the viscosity, η', is often called the dynamic viscosity.

Another form of dynamic experiment will be mentioned in passing which has been very useful in the investigation of biological materials. Using either a torsion pendulum or a beam of material mounted at one end only (see Ferry 1982 or Ward 1971) it is possible to subject the material to oscillations which decay freely with time. In such free decay the amplitude of the oscillations decreases exponentially with time. So if θ_n is the amplitude of the nth oscillation,

$$\frac{\theta_n}{\theta_{n+1}} = exp\delta, \text{ where } \Delta = \pi G''/G' = \pi\tan\delta \qquad [\text{Eq.1.33}]$$

Δ is also known as the *logarithmic decrement* and is an extremely useful experimental handle for the simple determination of G' and G'', especially at low frequencies up to about 10 Hz in a number of different test geometries.

1.5 SPECTRUM OF VISCOELASTIC BEHAVIOUR

It is now necessary to draw together the mathematical descriptions of viscoelasticity to show their interrelationships, how they can complement each other in the investigation of biomaterials, and how their results relate to the structure of the biomaterials at the molecular level. The response of a polymer to dynamic oscillations is probably the easiest to understand from the molecular point of view. The most basic variable with dynamic experiments is time – the frequency of oscillation. This is inversely proportional to time-after-loading in a transient experiment. The polymer molecules are in Brownian motion, just as described for rubber. The backbones are constantly changing their shape, rapidly at short range, but with the entire length of the molecular writhing more slowly representing a long-range average of the short-range motions. Any side groups are wagging and twisting. To a first approximation the proportion of molecular displacements which are in phase with the externally applied oscillations represents energy storage; the proportion which are out of phase represents energy dissipated as heat. The material can show the mechanical properties of a glass either if it is cooled (the amount of cooling required depends on the material and is typical of it) or by applying oscillations at such a high frequency that essentially no backbone motions occur during the period of oscillation. The effect of either of these treatments is to restrict the amount by which the molecule can respond to the external forces by changing its shape, and so the forces are concentrated on to the backbone of the molecule. Under these conditions the molecule behaves much as a Hookean solid.

Rubbery behaviour is typical of the plateau zone (Figs 1.11, 1.12). The polymer molecules, when excited at these intermediate frequencies, become entangled very easily, much as a ball of wool carelessly handled, and the entanglements act as labile cross-links, effectively transmitting the forces.

Between these two zones – the glassy and rubbery plateaux – exists a viscoelastic transition zone. As the frequency of imposed oscillation increases from the rubbery state, so the configurational changes in the network strands fail progressively to adjust themselves in the time allowed by the frequency of the oscillations. The long-range motions, being of lower frequency, are the first to run short of time in which to adjust, leaving shorter and shorter range motions to respond as the frequency rises. Gradually the strain in response to the applied stress diminishes and G' increases from the rubbery modulus of 1 MPa to the glassy modulus which is nearly 10 GPa.

What is happening to the loss modulus during these changes? G'' is a measure of the energy lost through 'viscous processes'. Relatively little energy is lost whilst the period of oscillation is not similar to the characteristic times which describe the rates of molecular processes involved in mechanical deformation. In the rubbery and glassy zones the oscillation period is different from these molecular resonances so losses, and G'', are relatively small. But in the transition zone the period of oscillation is similar to that of one or other of the molecular movements; the molecular movements lag behind the imposed oscillation dissipating large amounts of energy and giving a high loss modulus, thus contributing a greater viscous component. Obviously if there are several distinct molecular movements then there will be distinct discontinuities or secondary transitions. So the curves of G' and G'' will be rather more sinuous.

At the other end of the frequency range – the terminal zone of the modulus curve – entanglement slippage can occur within the period of oscillation and the molecules can assume any and all possible shapes. There is thus little restraint on the material and, if it is uncrosslinked, it will behave as a liquid of high viscosity. The terminal or flow zone will not appear if the material is crosslinked and the modulus recorded will be the equilibrium modulus of a stress-relaxation experiment: the relaxation modulus, $G(t)$, is approximately a mirror image of G' reflected in the vertical axis. The appearance of the zones is also affected by the molecular weight: if the molecular weight is low (below 10 000 d), the plateau zone is absent and the transition and terminal zones blend directly. Highly crystalline or glassy polymers will have a relatively high modulus over the entire frequency range, although there are still changes in the modulus which can give much information.

It will be remembered from the models of transient experiments that the relaxation and retardation times fall in the zone between the rubbery and the glassy states (Figs 1.16, 1.17, Eqs 1.24–1.29). The characteristic relaxation processes are the same as those occurring in the transition zone of dynamic experiments. In other words the relaxation times can be associated with the various modes of motion of the molecules. It is this basic association of molecular and mechanical properties which makes τ such an important and general constant, and which makes the relaxation spectrum, $H(\tau)$, such a useful form of comparison between tests whether transient or dynamic. Using the Alfrey approximation to derive $H(\tau)$ from $G(\tau)$, $H(\tau)$ can be derived from G' or G'' by the following relationship:

$$H(\tau) = \left[\frac{d\,G'}{d\,\ln\,\omega} \right]_{1/\omega=\tau} = \frac{2}{\pi} G''_{1/\omega=\tau} \qquad [\text{Eq.1.34}]$$

These transitions can be detected by another parameter, tan δ. Since G'' increases relative to G' in the transition regions, tan δ will also increase in these regions. Since G' and G'' are both increasing with frequency it is not very easy to compare them by eye. But the ratio between the two gives a much more delicate comparison, amplifying the differences and making them very obvious. Tan δ is therefore a much-used indicator of the presence, position and relative magnitude of transitions. As would be expected, τ and tan δ are closely related. For the Maxwell model, tan $\delta = 1/\omega\tau'$; for the Voigt model, tan $\delta = \omega\tau$. These relationships are, however, too simplified and formal to be of practical use in most instances.

So far temperature has not been mentioned except when referring to the glassy state. A polymer tends to the glassy state either as the temperature is reduced or as the experimental time gets shorter. Thus the high-frequency parts of the dynamic experiments and the first parts of transient experiments (the first fraction of a second – assuming that the loading is instantaneous) produce results equivalent to lowering the temperature in experiments with a longer time constant. Conversely higher temperature is equivalent to longer times in transient and dynamic experiments, bringing the polymer into a region of lower modulus. For this reason, application of heat to a glassy polymer (e.g. Perspex) softens or melts it, and the plastics moulding industry is made possible. This equivalence of time and temperature has been enshrined in the WLF (Williams, Landel, Ferry, the authors of the paper in which it was derived) equation of time–temperature equivalence or superposition. Its mathematics are beyond the present scope – see Ferry (1970)]. This relationship allows experiments performed at different temperatures with different time constants to be related to a continuous spectrum of response. There are several implications of this fact. The first is that although the different types of transient and dynamic tests are limited in the time ranges over which they are most effective, these ranges can be extended by the judicious variation of temperature. Although the practical range of temperature for biological materials is little more than from 0 to 40°C, even this can extend the time range by four to five orders of magnitude. Thus, although the experimentally convenient time scale for a transient test is about 10^0 to 10^3 s, use of the time–temperature interchangeability allows the range to be extended, from 10^{-2} to 10^5 or so. This would allow the analysis of a system with values of τ between 10^0 and 10^3 s. Thus the power of the transient experiment in practice becomes greater, which is a good thing because transient tests on the whole require less capital outlay in equipment. But a note of caution should be sounded. Time–temperature superposition theory has proved its usefulness with artificial polymers, both filled and unfilled (Section 5.5) but has not yet been adequately tested with the much more complex biological materials. In addition, time–temperature superposition is valid only when no new relaxation processes are made possible by the change in temperature. It is possible, for instance, that a particular relaxation process could occur only above a particular temperature as a result of chemical change due to temperature.

In addition, most biological materials are hydrated. The interactions of these materials with water will change with temperature (e.g. with elastin, Fig. 2.28) so the WLF equation cannot be used directly. The time–temperature relationship highlights another problem that has scarcely been touched upon. If a change in temperature, such as might be expected in nature, can change the reaction of a material to mechanical stimuli to the extent suggested above, then one might expect that mechanical tissues should be adapted to the temperature at which they function. Some evidence exists to suggest that this might be so with collagen and it seems to be true with elastin.

Another factor already mentioned, which is of great importance in biological materials, is water. Everyday experience says that water will swell and soften biological materials. This softening can be attributed to the increase of free space within the material allowing greater kinetic freedom for the polymer molecules. Another way of saying that the material has been softened is to say that G' and G'' have been shifted to higher frequencies or lower temperatures and that the brittle dry materials which become soft and pliable on wetting are being brought out of the glassy phase as the glass transition temperature is lowered by the addition of water, which acts as a diluent or plasticiser (Dorrington 1980; Wortmann 1987). The influence of water in proteins is much greater than merely affecting the glass transition temperature: the conformation of proteins is greatly influenced by the presence of a polar substance such as water (Section 2.3.4.2). Hydrophobic amino acids will tend to clump into zones which exclude the water, making the protein globular. Such hydrophobic interactions are largely beyond the scope of this book, but are of great importance in controlling conformation and molecular mobility (Sections 2.2, 2.3, 4, 5.3). In general it should be said that the role of water in the mechanics of biological materials has not received very much attention and could do with much more investigation. The presence of water is certainly very important.

1.6 YIELD AND FRACTURE

Two further aspects of the mechanical properties of polymers require some description: yield and fracture. Although not much has been done on the yield and fracture behaviour of biomaterials, this is an area where much more work may be expected, since one of the overriding characteristics of biological materials and structures is their toughness and resistance to rupture. Skin and wood are as tough as the best man-made materials, although only for wood do we have any idea why this should be so. Toughness is an important requirement for most biological materials – if bladders went pop with regularity that boilers do, we should all be in deep trouble.

There are two main ways in which a material can react when it is extended beyond its safe elastic limit. It can break immediately or it can undergo plastic deformation which is known as yield (in metals, ductility). Just as the mechanical

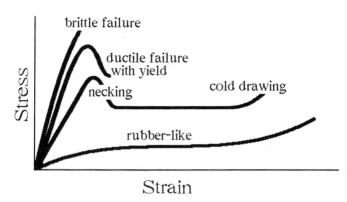

brittle failure

ductile failure
with yield

necking   cold drawing

rubber-like

Strain

Figure 1.21 Varieties of fracture and yield behaviour of a polymer at different temperatures: as the temperature increases so the initial modulus drops and the material stretches further.

properties of a viscoelastic material vary with temperature and strain rate, so do the yield and fracture phenomena (Fig. 1.21). Brittle failure is characterized by low strain and rupture occurring at the highest stress reached. Common brittle materials are biscuit, dessert jelly, high carbon steel and the membrane around a hen's egg just beneath the shell. Brittle materials are not common in nature. Also shown in Fig. 1.21 is ductile failure with yield just before failure. The yield involves plastic deformation. In the curve illustrating necking and cold drawing, the material has yielded which allow the cross section to be reduced quite abruptly. The material goes on extending with the polymer molecules reorientating themselves in the necked region. This process of 'cold drawing' leads to a material with its molecules in a preferred orientation which is much stiffer than the amorphous material from which it was derived, and which is therefore called 'strain hardened'. The strain hardening contributes to the stress--strain curve at point later on the curve when all the material has been cold drawn, leading to a final upturn before brittle failure. This shape of curve is also given by hair (Section 2.3.1) and the accepted explanation is somewhat similar. Rubbery behaviour is also illustrated for comparison. The phenomena of yield and plastic deformation are thus associated with molecular transitions and indeed the process of necking is a transition.

Interestingly enough most biological materials seem to have resistance to necking and yield built into their mechanics. This can be shown as follows: the true stress (Section 1.2) in a strained specimen, is higher than the engineering stress based on the cross-sectional area of the unstrained specimen. The true stress–strain curve can give the ultimate tensile stress using the Considère construction (Fig. 1.22). The tangent to the stress–strain curve drawn from (-1) on the strain axis gives the maximum stiffness of the material. Beyond the point at which the tangent touches the curve the true stress is dropping and the material is failing. Figure 1.22 show this for a material with a convex stress–strain curve.

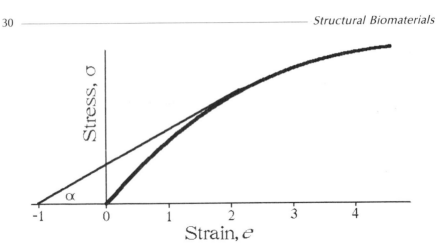

Figure 1.22 The Considère construction as a criterion of liability to yield (Eq.1.32).

But nearly all biological materials have a concave stress–strain curve to which such a tangent cannot be drawn. So there is no possibility of yield within the working strains of the material, which would not be so were the stress–strain curve more like that of rubber (Fig. 1.23). This means that even if the materials are working near or in their transition zone the energy fed into the material as it is extended will be spread evenly throughout, and there will be little possibility of local increases in stress which, as shall be explained, can lead to the failure of the material at low overall loads.

The fracture behaviour of both polymers and the more complex biomaterials is relatively unknown and unexplored, especially since this usually occurs at, mathematically difficult, high strains. The fracture of brittle materials is much better understood, the foundations being laid by the theory of Griffith (1921). Although the theory is not difficult, it is little applicable to polymers in the viscoelastic and rubbery regions of their behaviour. Essentially, Griffith said

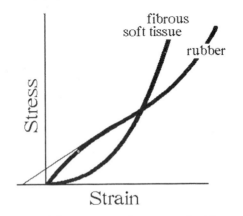

Figure 1.23 Comparison of the stress–strain curves of rubber and a typical fibrous soft biological tissue.

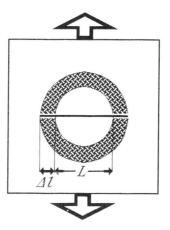

Figure 1.24 Model for deriving the general conditions for the propagation of cracks.

that a fracture results in the formation of two new surfaces on each side of the crack and that the formation of these surfaces requires energy. This energy is stored as changes in bond length throughout the rest of the material as it is stretched. The process of fracture then involves the transmission of this energy to the fracture surfaces, at the same time relaxing the strain in the area from which energy has been released. If the crack is considered to be linear, travelling at right angles to the direction of the applied stress (Fig. 1.24), then it is reasonable to suppose that the energy which the crack is absorbing comes from an area defined by the crack as the diameter of a circle. It is fairly obvious that as the crack (length L) extends (by an amount Δl), the amount of energy available

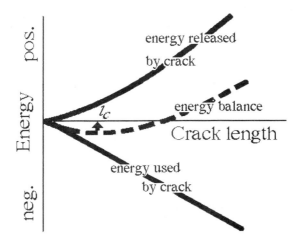

Figure 1.25 The energy conditions associated with the propagation of cracks showing the derivation of the critical crack length, l_c.

for the propagation of the crack will increase at a greater rate since it is proportional to the square of the crack length. Up to a certain crack length the energy released from this area is not enough to propagate the crack, but after this point (the critical or Griffith length) more energy is released than is required so the crack is propagated (Fig. 1.25). This mechanism, plus the capacity to transmit the stresses to the crack tip where fracture is actually occurring, will account for the fracture of brittle (= non-ductile) materials. But ductility or plasticity (i.e. irrecoverable deformation) can use up energy before it can get to the crack tip to contribute to the new fracture surface. In metals, ductility can account for far more energy than is needed for propagation of the crack and this has also been shown to be so for brittle fracture in polymers. In more compliant polymers and complex materials such as skin, this is expressed in motions such as the alignment of the polymer molecules ahead of the crack tip.

Arising largely out of Griffith's work is the realization that two criteria have to be fulfilled for a piece of material to break. The strength of the material must be reached and there must be sufficient elastic strain energy available at the tip of the crack to propagate it.

In most of the attempts to produce a theory of fracture, one of the aims has been to produce figures which are material parameters and which do not depend upon the shape of the specimen or the orientation of the crack. This is a reasonable aim which has substantially been achieved for materials which are linear, in which all the strain energy is involved in fracture and in which fracture occurs at relatively small strains. But one of the properties of polymers at large strains which has already been mentioned is the tendency for the molecules to become orientated in response to the deforming forces. This can, in some materials, lead to strain crystallization somewhat akin to the strain hardening of a cold-drawn polymer. Such crystallization can be reversible and so be a function of strain. So for such materials the problem arises that its morphology is changing with strain. And just as the strain hardened polymer has different properties from the random polymer, so the properties of any polymer may be considered to change with extension. In fact a cross-linked polymer in a state of large static strain, at equilibrium, may be considered as a new anisotropic material whose linear viscoelastic properties could be studied (Ferry 1970). So it should be no surprise that there is (probably) no such thing as a unique work of fracture for high-strain polymers. The other complication is that energy can be dissipated at sites remote from the fracture surface. (Oliver Wendell Holmes' 'The Deacon's masterpiece; or the wonderful "one-hoss shay", a logical story' (Holmes 1907) shows this to perfection. All the knocks of everyday life were accumulated within the perfectly balanced and adjusted structure and kept away from any fracture surface until, exactly a hundred years after it was made, the shay collapsed into a heap of dust because it could absorb no more strain energy. Some fibre-glass car bodies were rather like this.) Some indication of this dissipation can be obtained from hysteresis tests, the hysteresis loss being energy lost within the bulk of the material. Should one take account of this loss when calcuiating the fracture constants? For filled rubbers, in which the amount of filler (carbon black) is varied, it is found that not only is this directly correlated with the degree of hysteresis but that higher hysteretic losses go with greater toughness. The problem has not yet been investigated, but it may well be that mechanisms for energy dissipation also vary with strain. Still a further factor is the transmission of strain energy to the crack tip where it is needed to supply the energy for crack propagation. The effectiveness of the transmission of stresses to the developing crack tip is a function of shear modulus, which itself changes with shear rate. So the speed at which the crack is fed with energy will depend upon the rate of change of shape in the material around the crack.

Toughness can be increased by a number of mechanisms, all of which increase the amount of energy required for fracture and all of which can be present in a tough material. Here are some of them:

1. The strain energy is unable to reach the crack tip. For instance it can be dissipated by plastic yield and failure of the material remote from the crack. It is quite possible that viscous effects within the material will slow down the rate of delivery of energy to the crack tip so that the crack can be propagated only slowly and with difficulty. Transfer of fluid from one site to another within the material comes into this category, and seems to be a mechanism for toughening teeth (Fox 1980) and very probably other biomaterials. The strain energy may not be transmitted at all if the shear stiffness of the matrix material is too low (evidenced by a J-shaped stress–strain curve, common in soft tissues – Section 4.2) (Mai & Atkins 1989);

2. The total energy required for cracking is raised, e.g. the fracture surface very convoluted and therefore of large area or the material at the crack tip deforms plastically;

3. The stress at the crack tip is de-focused by, for example, increasing its radius of curvature or by the Cook-Gordon effect (Fig. 5.1); The sharpness of the crack tip governs the stress intensity. It focuses the strain energy onto the next susceptible bond. At high strains in unidirectional extension, the crack tip rounds off into a semi-circle. In rubbers (and probably other high strain materials) there is evidence of strain crystallization at the crack tip (now a semi-circle) which will further strengthen this most vulnerable area.

4. As the crack opens, fibres or filaments extend across it dissipating energy by their own deformation or by friction as they pull out from the bulk of the material.

5. The material is prestressed in the opposite sense to that in which it is most likely to be loaded (e.g. in compression if the most likely loads will be tensile) so that a crack cannot start until this prestress is paid off.

6. The entire structure is so small that the strain energy necessary for fracture cannot be stored.

In practical terms, fracture toughness can be measured and calculated in a large number of different ways. For each test geometry there is a specific mathematical solution which makes a number of assumptions about the material

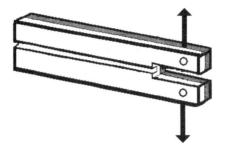

Figure 1.26 Notched double cantilever beam test piece used for the derivation of the data in Fig. 1.27.

and the test and allows calculation of toughness from a number of more or less simple measurements. General information is available from Atkins & Mai (1985). However, biological materials frequently transgress these assumptions, being anisotropic or very stretchy or inhomogeneous or of an odd shape. There is a pragmatic way of coping with these problems: use a test in which the crack grows in a stable fashion such that the test piece can be unloaded (i.e. returned to its original length or shape, or until the recording device shows that no load is still being applied) before the test piece has broken into two pieces. An example of this is the double cantilever beam (Fig. 1.26). If a sample of this shape is loaded as indicated, it will, when it reaches breaking load, start to fracture (Fig. 1.27). If the fracture is allowed to propagate with increasing strain and the test piece unloaded before the specimen has broken in two, the force--extension curve will follow the path MNO. Notice that this is not a stress–strain curve: the lines MO and NO differ in slope because the cross-section of the test piece has been reduced by the cracking. Notice also that because the test piece has been unloaded and returned to zero load, the area within the triangle OMN represents the work done in fracturing the specimen. In a viscoelastic or elasto-plastic material there are additional losses due to hysteresis or permanent deformation on loading

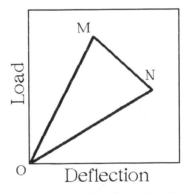

Figure 1.27 Load–deflection curve for the estimation of fracture toughness.

and unloading. Corrections can be made graphically (Jackson *et al.* 1988). This type of test can be performed with morphologies other than the double cantilever beam: so long as the crack propagates in a stable fashion and can be arrested at the will of the experimenter, it will give results which can be analysed in this fashion. Atkins & Mai (1985) give the criteria for controlled cracking in a number of morphologies.

Alternatively, if the test is performed at a sufficiently slow rate of extension such that the force has fallen to zero just as the crack has travelled across the specimen and it has completely broken, then there can be no strain energy left in the material and the area under the curve represents only the energy used for fracture. This is a rather risky trick and implies that you can talk to your sample in its own language! Either way, any elastic strain energy is discounted from the final reckoning, and the energy which the force–deformation curve encloses is that required to propagate the crack. The great advantage of this graphical technique is that it is entirely independent of any mathematical model and the assumptions involved in generating such a model. As such it is particularly useful for biological materials, which are so complex that there frequently isn't a respectable mathematical theory to describe or analyse their fracture processes.

There is another problem. Any piece of tissue will have a number of imperfections (scratches, nicks, notches and cuts) whose size, nature and distribution are difficult to control or predict. Depending on the nature of the material, these imperfections can affect, or even direct, the mechanisms of failure. For instance, they can initiate a crack. This is because any imperfection will have the effect of concentrating stress around it, more especially at sharp corners (Gordon 1976). Smooth corners and edges are important in controlling fracture. One strategy is to confine the deformation to a very small area, effectively limiting failure to a small zone. This is achieved in the 'trouser-tear' test (Section 4.6) and by techniques involving cutting or wedging. A third strategy is to introduce an imperfection larger than any of those already in the test piece. This is commonly done by notching or cutting the specimen (Section 5.2.2).

The problems involved in fracture of biological materials when they are stretched in two directions at right angles to each other (bi-axial straining – the above discussion has been concerned with uniaxial straining only) has not been investigated yet, even though it is of greater relevance to organisms. It seems likely that fracture of a biaxially strained specimen will be more 'brittle' than fracture of a uniaxially strained specimen. Compare the way in which a balloon pops when pricked with a pin with the way a piece of rubber from the same balloon reacts to the same stimulus when stretched by the same amount but uniaxially. The 'toughness' of rubber is dependent upon the way in which that toughness is measured. So it is quite possible that the toughness of skin will vary with its position on the body in accordance with the direction and magnitude of likely strains.

1.7 ADHESION

Adhesion will be mentioned only briefly in that the proper tools for its measurement have been discussed in general terms already. There are a number of different mechanisms of adhesion (Kinlock 1987) which can be summarized as mechanical interlocking (micrometre-sized roughness); diffusion of one component into the other (has been observed when two artificial rubbers are pressed together); electron transfer (an arcane and mostly insignificant theory and effect) and adsorption (by van der Waals forces, hydrogen or chemical bonds – see Fig. 2.7). In general the strength of an adhesive bond is probably best thought of as a problem in fracture, so it can be measured with the caveats of fracture mechanics in mind. Gordon (1976) gives a good discussion of adhesive joints – both their formation and strength or toughness – in his treatment of how wooden aircraft are built and maintained.

CHAPTER 2

Proteins

Proteins are polymers of amino acids; polysaccharides are polymers of sugars. Between them these two groups of substances make up nearly all the skeletal tissues, pliant and stiff, of animals and plants. The precise complement of amino acids or sugars and the order in which they arranged along the chain ultimately controls the mechanical properties of the material which they form. These chemicals and the materials they form are the subject of this and the next three chapters.

Proteins and polysaccharides of mechanical significance can be divided into two main groups – fibrous and space-filling. In general, proteins are more important as fibres, represented mainly by collagen. Polysaccharides are just as capable of forming fibrous materials, but their major function seems to be stabilizing water in many biomaterials. The reasons for this dichotomy of function are not clear, and there are probably many. It may be to do with the ease with which hydrophobic interactions – important in stabilizing interactions in an aqueous environment – can be formed by the two classes of polymer; it may be to do with the diversity of bonds which hold the monomer units together. This is a distinctive difference between proteins and polysaccharides, since polysaccharides can be linked in many more ways than amino acids and can form branching polymers. There may be other important reasons for the dichotomy of function.

An important general aspect of both groups of polymer is the way the units are bonded together and the way they interact with each other, both as neighbours in a polymer chain and as interacting lengths of polymer. Ultimately it is these interactions which will dictate the reaction of the polymer to mechanical stress in a given environment. Many mechanical tests designed to probe the nature of such materials do so by varying the conditions of temperature, humidity or chemistry around the bonds and seeing how the response to mechanical stress varies.

2.1 Amino acids and their polymerization – primary structure

One way of approaching the mechanical properties of proteins is to consider the bond-forming and space-filling properties of the amino acids, then to see

37

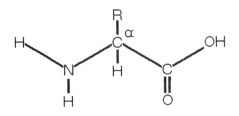

Figure 2.1 The basic structure of an amino acid. Commonly found side-groups (*R*) are shown in Fig. 2.6.

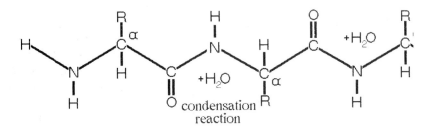

Figure 2.2 Polymerization (condensation) of two amino acids.

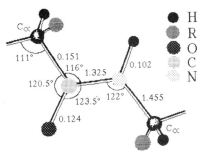

Figure 2.3 The basic dimensions (in nm) of the peptide bond.

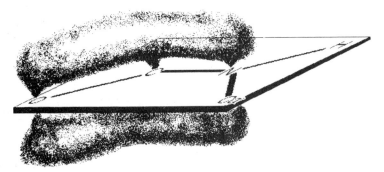

Figure 2.4 The electron cloud around a peptide link, holding the amide group in a single plane thus restricting rotation around the backbone.

what structures they take part in and why (Dickerson & Geis 1969, Richardson 1981). The basic structure of an amino acid is shown in Fig. 2.1. The central carbon atom is called the α-carbon; R may be one of many possibilities – the commonest are shown in Fig. 2.6. Amino acids are joined together or polymerized by the peptide bond or amide link which is formed during protein synthesis in the ribosome as a condensation reaction (Fig. 2.2). The angles and dimensions of the peptide bond are shown in Fig. 2.3. Although Fig. 2.2 shows a double bond between the carbon and the oxygen and a single bond between the carbon and the nitrogen (the amide bond itself), in reality the two average out to give a low-energy double bond arcing from the oxygen to the carbon to the nitrogen (Fig. 2.4). The effect of this is to hold the amide group in a single plane, an almost constant form for the peptide bond to take. This means that almost no rotation is possible about the peptide bond, but there is freedom of rotation about the single bonds between the amide groups and the single α-carbon atoms. This allows the polypeptide chain to assume a large number of configurations. The actual number of configurations possible can be shown to be limited by further considerations. These are basically:

1. The shapes of the backbone allowed by the limited rotations about the bonds to the α-carbons.
2. The limitations on (1) set by the size of the side-group, R.
3. The differing types of stabilizing secondary bonds (notably hydrogen bonds) formed between various oxygens and hydrogens (notably those associated with the peptide bond).
4. The chemical nature of the side chains and the interactions of these groups with each other and with the immediate chemical environment.

The polypeptide chain has two degrees of rotational freedom per residue: the twist about the α-C-N bond axis (ϕ) and the twist about the α-C-C axis (ψ). A list of all the (ϕ,ψ) values for all the residues will completely define the shape of the chain. The permissible values for ϕ and ψ can be calculated to lie within certain bounds defined by the minimum distance to which the unbonded atoms of the peptide bond can approach. Some values of ϕ and ψ are not permitted because the unbonded atoms of neighbouring groups would approach too closely. If the values of ϕ and ψ at every α-carbon are the same, the chain naturally falls into a helical shape. The number of amino acids per turn and the distance between the turns (the pitch of the helix) are both determined once ϕ and ψ are specified. Obviously parameters of such general importance as ϕ and ψ, whose values can be derived from purely theoretical considerations, can be used to show what regular shapes or conformations of the peptide chain are possible. This was done by Ramachandran and his colleagues at Madras, and a conformation map or 'Ramachandran plot' for the polypeptide chain is shown in Fig. 2.5. The region from $\phi = 20°$ to $\phi = 140°$ is particularly favourable. This corresponds to a rotation of the N–H bond toward the side chain and the removal of the CO group of the same peptide as far as possible away from the side chain. Within this region (marked) are

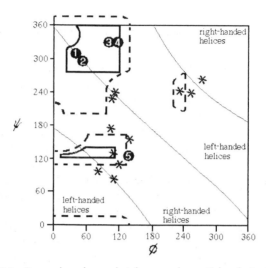

Figure 2.5 Ramachandran plot for a polypeptide chain (see text).

the anti-parallel (1) and parallel (2) β-pleated sheets, the polyproline helix (3), the collagen coiled coil (4) and the α-helix (5) (numbers refer to figure 2.5). Other helices and ribbons can be plotted on this diagram (asterisks). These various conformations are secondary structures, the primary structure being the sequence of amino acids. Three major points must be made about these regular structures. They are stabilized by interactions between the hydrogens of the NH groups and the oxygen of the $C = O$ group. These hydrogen bonds are of great importance in the selection of the most stable conformation. Several helices are possible with hydrogen bonding occurring between the $C = O$ of one peptide link and the NH of the second, third or fourth successive peptide link. The most favoured helix – the α-helix – has the hydrogen bonding between one peptide link and the third following link, making the hydrogen bonds particularly unstrained and stable. This helix is the most stable and commonly found. Such helices (and other regular structures) have a stability which is a function of their size – i.e. the number of residues contributing to the structure. In the α-helix, for example, the protein chain, originally kinetically free, is arranged to form one turn of the helix allowing the formation of one H-bond. But with two turns of the helix, four H-bonds are formed; with three turns, eight H-bonds are formed; with n turns ($3.6n-2.6$) (or the integer just below this figure), H-bonds are formed. So the longer the helix, the more stable it becomes due to the cooperative interaction of many local forces to achieve long-range order. This is obviously only a first approximation as will be apparent when other interactions involved in the formation and stabilization of secondary structure are considered. Finally, the side-group can affect both the rotational freedom of the bonds in the peptide link and the closeness of packing of the amino acids.

2.2 CONFORMATION – SECONDARY STRUCTURE

Computer analysis of the conformational areas of proteins and amino acids comprising the different conformations shows that close correlations can be made between conformation and the occurrence and position of specific amino acids. These correlations are being studied in model systems of polypeptides made with a single type of amino acid. The conformational correlations obtained from the study of native proteins, which are far more complex, seem mainly to be confirmed by these studies. The implications are then threefold: first, the control of protein conformation must be largely a localized affair. If there were a significant contribution from more remote amino acids, as might occur if two parts of the protein were folded close to each other, then such a good correlation would not be possible. Second, the protein must assume its final configuration irrespective of the conformations which it adopts beforehand (e.g. during synthesis or some renaturation process). But if each individual residue of an unfolded chain could exist in only two states (a gross underestimate) then for a chain of 150 residues there would be 10^{45} possible conformations. If each conformation could be assumed with the frequency of a molecular rotation $(10^{12}\,s^{-1})$ then it would take about 10^{26} years for all possible conformations to be assumed. It takes about two minutes for a protein such as lysozyme to be synthesized and to assume its native conformation, so there must be a limited number of conformational pathways for the protein to take towards the assumption of its native conformation, which will be a conformation with relatively low free energy. The sort of determinism of biology, which underpins the logic of the subject, thus states that evolution (with thermodynamics dictating the folding) has selected the amino acid sequence to form a biologically useful molecule, with presumably a limited number of pathways from the unfolded form to a unique native structure of low free energy. All this must be taken to be true for a single type of environment of the protein (e.g. aqueous with a particular pH, salt concentration, etc.) so that a protein can be secreted, by a cell, in one conformation, acted upon by a different environment outside the cell, and assume another conformation. This is the basis of so-called 'self-assembly' systems of the extracellular matrix. Cellular products can be designed to have specific and changing reactions with other extracellular components and thus build up tendon, bone, cuticle and the like. The third implication of these correlations is that it is possible to predict something of the conformation of a protein simply from its amino acid sequence. It has thus been found possible to assign amino acids to various categories according to their helix-forming tendencies (Table 2.1). There are several factors that determine the conformational preference of a given amino acid residue. For instance the most likely state for a protein chain in the absence of any stabilizing factors is irregular. This is the conformation of highest entropy (Section 1.2). So an amino acid which does not have any factors which are favourable to the formation of

Table 2.1 The helix forming properties of amino acids

Helis breaker	Helix indifferent	Helix former
Glycine	Lysine	Valine
Serine	Tyrosine	Glutamine
Proline	Threonine	Isoleucine
Asparagine	Arginine	Histidine
Aspartic acid	Cystine/cysteine	Alanine
	Phenylalanine	Tryptophan
		Methionine
		Leucine
		Glutamic acid

α-helices must necessarily be a helix breaker. This argument seems to hold for glycyl residues. They have no side chain and so have no energetic factors favourable to the formation of helices: thus the entropy of irregular zones makes glycyl residues helix-breaking. When a β-CH_2 group is added (giving alanine) the resulting possible interactions favour helix formation. But if a $CONH_2$ group is now added, to give asparagine, the polar side chain interacts with the amide group of the peptide link and the helical conformation is destabilized. These electrostatic effects are less if the polar group is further from the peptide link. Thus the insertion of an extra CH_2 group into the side chain to give glutamine results in a helix-forming residue. A similar argument applies for the series aspartic acid–glutamic acid. Aspartic acid has a charged group which can form H-bonds with neighbouring backbone amide groups when in the random conformation so is a helix breaker. Glutamic acid has the same charged group spaced further from the backbone by a CH_2 group which thus cannot take part in such interactions. Serine is a helix breaker because it has a charged group close to the backbone, but proline is a helix breaker because the rotational freedom of the $-C-N$ bond is severely restricted by the ring structure. Proline (and the related hydroxyproline) is extremely important as a constituent of collagen.

There are further considerations in the assumption and stabilization of particular conformations in proteins. The chemical nature (acidic, basic, polar, non-polar, etc.) of the amino acids is important in determining medium and long-range interactions. A medium-range interaction may be said to extend from one residue to a distance of (say) four residues on either side. The α-helix is thus stabilized by medium-range interactions in the form of hydrogen bonds. Long-range interactions involve other amino acids and the environment. In an aqueous environment polar side-groups will tend to interact with the water to a much greater extent than the non-polar groups, tending to keep the polar groups to the outside of the protein and the non-polar groups directed towards the centre. Non-polar interactions also seem to be very important in stabilizing interactions between neighbouring proteins, in effect forming 'sticky' areas on the surface of proteins.

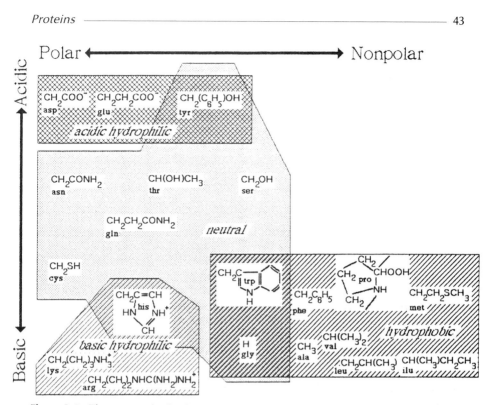

Figure 2.6 The range of side chains found most commonly on amino acids, grouped according to broad chemical properties.

Side chains (Fig. 2.6) will interact with each other to form salt linkages or electrostatic bonds (between ionic groups of opposite charge); hydrogen bonds (between polar groups); van der Waals and London forces (between non-polar groups); covalent disulphide links (between sulphur-containing side chains). The strength of these interactions is important when considering the mechanical properties of the proteins in which they occur since it is largely these interactions which give the material its integrity and are resisting the externally applied mechanical forces. Figure 2.7 shows the relative strengths of these bonds. It must

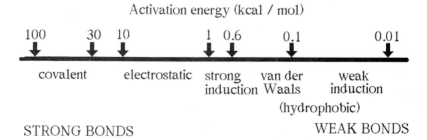

Figure 2.7 The relative strengths of inter-molecular interactions.

also be emphasized that not only are there more types of structure than the trilogy of α-helix, irregular and β-sheet (e.g. tight or β turns, collagen helix,) but that it is highly likely that many proteins which are labelled irregular will be found to have regions of well-defined conformation. This problem is illustrated by the β turn which was the first non-repetitive structure to be found, and by elastin which relies on specific conformational changes for its elasticity, yet shows many of the attributes of a rubber, which is the archetypal random structure.

The mechanical properties of biological materials should, ideally, be explicable in terms of their molecular structure and interactions. Although biological materials are much less well understood than the simpler organic polymers on which viscoelasticity theory is based, it is still possible to make some broad generalizations and even to show how quite subtle molecular changes can be related to mechanical properties.

2.3 STRUCTURAL PROTEINS

There are several families of structural proteins: keratins, collagens, silks, arthropod cuticle matrices and elastins among them. These families are of differing complexities and are based on chemical and morphological rather than molecular and mechanical criteria. Thus the keratins, typified as being sulphur-containing proteins (Section 2.3.1), can have α-helical, β-sheet or irregular conformations. Collagens have their own typical conformation but some silks contain collagen as does mussel byssus thread, technically a scleroprotein in some ways more closely related to insect cuticle. Silks also show all the conformations shown by keratin and more besides. The mechanical properties of these various materials are much more closely related to the molecular mechanisms than to the overall chemistry or morphology on which their classification is based. In what follows, more significance should be given to these mechanisms than to the particular class of material from which the examples are taken.

2.3.1. Keratins

Keratins form a group of varied proteins, mostly found in vertebrate animals, which contain significant amounts of sulphur or tyrosine-based cross-linking which stabilizes the material (Fraser & MacRae 1980, Bereiter-Hahn *et al.* 1984). It is this sulphur which gives the nasty smell when horn, hair, hoof, feather, skin and the like are burned. Probably many different types of keratin exist, but they can be broadly divided into mammalian, avian and other (e.g. reptilian). In mammals, keratin occurs in skin, hair, horn and hoof; it has been most studied in hair, more especially wool (Fig. 2.8). Early work by Astbury, using X-ray crystallography (Astbury & Woods 1933) showed that the major part of

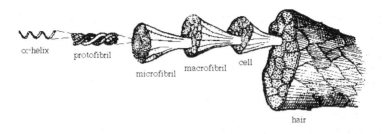

Figure 2.8 The hierarchical structure of α-keratin in hair (a hairarchy?).

hair keratin exists in one form (α) in the relaxed state and in another form (β) when the hair is extended in steam and held extended. Later it became apparent that the α form is the α-helix and the β form is the anti-parallel β-sheet which results from pulling the α-helices beyond their yield point. The steam provides the heat to break the H-bonds stabilising the helix: most H-bonds are disrupted at temperatures from 40 to 60°C depending on their environment. However, the α-helices are only the smallest component of a hierarchical system. They are assembled into larger units forming rope-like structures (Fraser *et al.* 1976) or protofibrils. Electron microscopy shows that these are embedded in a non-fibrous

Table 2.2 Amino acid contents of some keratins

Amino acid	High sulphur proteins			Low sulphur proteins		
	Wool	Horn	Hoof	Wool	Horn	Hoof
Lys	0.6	1.0	1.0	4.1	4.1	5.0
His	0.7	1.0	0.9	0.6	0.8	0.8
Arg	6.2	5.4	6.0	7.9	7.8	7.1
Asp	2.3	4.7	4.3	9.6	8.9	10.1
Thr	10.2	9.6	10.2	4.8	4.9	4.4
Ser	13.2	11.2	11.8	8.1	8.4	7.8
Glu	7.9	6.1	6.9	16.9	15.8	17.7
Pro	12.6	12.4	13.0	3.3	3.6	2.4
Gly	6.2	9.0	7.2	5.2	6.9	6.3
Ala	2.9	3.2	3.2	7.7	7.4	7.2
Cys½	22.1	16.3	16.9	6.0	4.7	3.7
Val	5.3	5.6	5.8	6.4	6.2	5.9
Met	0	0	0	0.6	0.7	0.7
Ile	2.6	3.3	3.2	3.8	3.8	3.9
Leu	3.4	5.3	4.9	10.2	10.1	11.1
Tyr	2.1	3.3	2.3	2.7	3.3	3.2
Phe	1.6	2.6	2.4	2.0	2.5	2.0
Helix breaker	34.4	37.3	36.3	24.3	27.8	26.6
Helix former	22.8	24.5	24.9	46.1	44.8	47.3
Helix indifferent	42.7	37.9	38.9	27.5	27.3	25.4
Percentage sulphur (total)	3.75	2.13	2.15			

From Marshall & Gillespie (1977)

matrix; fibre and matrix together constitute a two-phase composite material, a common form of construction for biological materials (Section 5.2). The matrix seems to be mainly composed of a heterogeneous assemblage of proteins containing much cystine. Table 2.2 shows the amino acid analysis of groups of these proteins. The high sulphur proteins of wool not only contain over 20% cystine (mouse hair contains up to 35%) but also significant amounts of proline and serine. Thus it is not surprising that the matrix which these proteins form should be considered to be in a relatively non-crystalline state and certainly not α-helical with such a high content of helix breakers. These proteins are also highly cross-linked with sulphur bonds, modern evidence suggesting that these bonds are intra- rather than inter-molecular. The high sulphur proteins of horn and hoof have as high a content of helix breakers as wool but only two thirds the content of cystine. Horn and hoof probably therefore have a much lower density of S–S links in the matrix. The low-sulphur proteins of all three materials are similar and can be correlated with the fibrous helical phase. But the presence of proline and cystine in this fraction suggests that these proteins may have a non-helical component which is cross-linked into the matrix. This leads to a fringed micelle model for the fibrous phase, a model which is supported by evidence from other sources such as partial sequencing of the amino acids which suggests that the proteins form distinct functional lengths of helical and non-helical conformation. Fringed micelle arrangements (Fig. 2.9) are probably very common in biological materials.

The mechanical behaviour of horn and hair can largely be explained in terms of a two-phase or composite model (see Chapter 5). When wool is tested in differing humidities with differing water contents (Fig. 2.10; Hearle *et al.* 1971) the initial Hookean part of the curve has the same slope for all the samples and gives a modulus of about 4 GPa. The yield point, at a strain of about 0.02, marks a change in modulus which drops by a factor of ten or so. Such a yield or transition indicates a change in the molecular structure. The most likely explanation, and one which is supported by evidence from X-ray diffraction, mechanical tests and molecular modelling, is that the H-bonds which stabilize

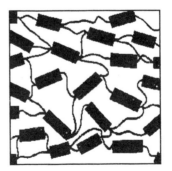

Figure 2.9 Fringed micelle model.

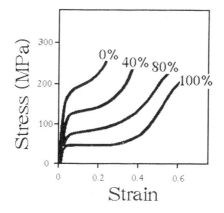

Figure 2.10 Mechanical properties of α-keratin as typified by wool at different relative humidities (Hearle *et al.* 1971).

the helical structure rupture and the helices start to unravel and form β-sheet structures. Experiment shows the modulus to be increasing after a strain of about 0.3 (in the wet) at which only about a third of the α-helical material has been unravelled (Danilatos & Feughelman 1979). Obviously there must be other constraints. The influence of water is also of great interest. The tendency of water (acting as a plasticizer – see below) is to reduce the sustained stresses in all parts of the curve except for the initial modulus. As water enters the hair, axial swelling is small (about 1% at 100% relative humidity) and the ratio between Young's modulus and the shear modulus changes from 2.7:1 (indicative of a more-or-less isotropic material) to about 0.5:1 (indicating a highly anisotropic material). Indeed using Eq.4.1, the respective Poisson's ratios are 0.35 (quite reasonable for a stiff material at low strains) and 0.9, (Fraser & MacRae 1980) which is much nearer that found for some soft tissues (Section 4.3) which consist of uniaxially orientated fibres in a pliant phase. The inference is that water is not entering the helical portion but is entering the matrix phase. Another aspect of water absorption is that the swelling is too great for the matrix to be a rubbery material in extended random coil conformation. It is probable that the matrix is a globular material with a non-random, but not fibrous, structure. Which implies that the matrix has a more or less determinate structure. The post-yield phase has also been shown to involve exchange of S–S bonds within the matrix, so that whereas the fibrous phase seems to be relatively stable and crystalline, the matrix phase seems to be labile and amorphous. The biological significance of all this, at least for hair, is problematical. It is difficult to think of any sequence of events in which the α to β transition could have any biological significance. The same may not be true for horn and hoof.

Rhinoceros horn has a similar stress–strain curve to that for hair when tested along the fibre, but a much lower modulus, with no yield, when strained at 90° to the fibre when presumably only the matrix proteins are taking the load

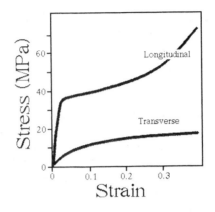

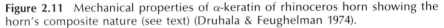

Figure 2.11 Mechanical properties of α-keratin of rhinoceros horn showing the horn's composite nature (see text) (Druhala & Feughelman 1974).

(Fig. 2.11; see also mechanical analysis of insect cuticle). It is therefore mechanically, as well as chemically, similar to hair, but matrix and fibre mechanics can be investigated independently by choosing the orientation of the test piece. However the stiffness of the helices themselves must be greater than the stiffness of the material as a whole, since the helices occupy only a fraction (in antelope horn, about 3/5) of the total volume. Application of a fibre composite model to results from mechanical tests on horn keratin (Section 5.2.3) gave a figure of 6.1 GPa for the stiffness of the fibres, which have an effective length of about 40 nm. The stiffness of the matrix protein varied from 6.1 GPa (dry) to 0.9 GPa (wet), although a slightly different technique gives a value of 1.2 GPa for the stiffness of the α-helix in wool. A higher value approaching 10 GPa has been measured in more or less perfect arrays of α-helices (polyglutamate) obtained by orientating them on the surface of a Langmuir-Blodgett trough (Wegner 1989). The wet matrix contributes little to the

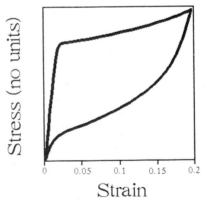

Figure 2.12 Hysteresis of α-keratin in hair.

post-yield modulus (Fig. 2.11). The matrix seems to be a more labile phase and as such is probably the basis of the hair-styling industry: under the influence of high temperature and humidity the H-bonds break leaving the hair more pliant. The hair can then be held in the chosen position while the temperature and humidity return to ambient allowing the H-bonds to reform in new positions. This holds the hair in its new, 'permanently' waved shape.

If hair is extended and then allowed to retract, it is found to show very high hysteresis (Fig. 2.12). This experiment seems not to have been performed with horn or hoof, but it seems very likely that it would give the same result since the classical interpretation of the curve is that on the return part of the curve the α-helices are reforming, relying on the matrix to provide a large part of the elasticity. Hysteresis, though not of this magnitude, has also been found in rubbers filled with finely divided carbon black particles (Section 3.2). It seems also to be intimately associated with toughness in these materials. The argument is that energy used up in processes with high hysteretic losses is unavailable for fracture. Thus all α-keratins are probably very tough materials. Whilst it is difficult to see what importance this might have for hair, it is clearly of use for horn, hoof and the outer layer of skin (stratum corneum) which have to be very tough. The morphology of horn keratin is relatively simple – it seems to be relatively easy to model (Section 5.2.3) and has a work of fracture of about $10 \, kJ \, m^{-2}$ (Kitchener 1987). This is a typical value for a high-performance composite (Harris 1980). The keratin of hoof and baleen is much more complex and hierarchical, being composed of spirally wound tubes of keratin in a keratin matrix. Wet hoof keratin is relatively soft with a Young's modulus of only about 0.5 GPa (Bertram & Gosline 1986). For a material which is galloped upon, used for digging and fighting and which cannot be instantly replaced if it breaks, toughness is obviously important.

Stratum corneum, the tough outer layer of the skin, is a rather more diffuse α-keratin with the fibres orientated more or less randomly. This is reasonable if the material is to accommodate stresses from several directions. This diffuseness is reflected in the tensile test curves (Fig. 2.13; Papir *et al.* 1975) which show that although there is a yield point, it does not separate an initial region whose modulus is independent of water content from a post-yield region which is dependent on water content. This suggests that the matrix is making an important contribution to the initial modulus as well as to the post-yield modulus. The way the modulus changes with water content shows a marked change at 70% relative humidity (Fig. 2.13) which could be called the onset of plasticity. This change is associated with a change in the way that water is bound into the matrix, as shown by differential scanning calorimetry (DSC). This technique measures the heat exchanges as the temperature of the sample is raised at a constant rate. If the specimen is cooled in liquid nitrogen so that all the water in it is frozen and it is totally glassy, then as the specimen is warmed up any frozen water that melts will absorb a latent heat of fusion as it does so, producing a peak (endotherm) on the trace of heat absorption against time

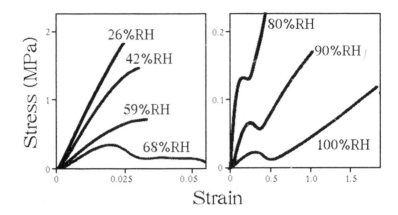

Figure 2.13 The effect of sorbed water on the mechanical properties of stratum corneum from the skin of a newborn rat (note changes in strain scale) (Papir *et al.* 1975).

which is the output of the DSC. If the water is closely associated with the protein (often called 'bound' water) then it is less likely that it can form ice with unstrained bonds. The strained bonds in the ice that does form from this water will be less stable than unstrained bonds and will be destabilized at a lower temperature than those of ice formed from free water. Thus 'bound' water tends to have a lower freezing/melting temperature than free water and this can be detected by DSC. The amount of heat absorbed by the melting water gives an estimate of how much water is present in that 'bound' fraction. With stratum corneum, DSC studies in conjunction with mechanical tests suggest that the onset of plasticity is associated with a particular fraction of the 'bound' water - - i.e. water which is immobilized within the matrix in a particular way. This could be due to at least two effects: the interpolation of water molecules into H-bonded linkages between the amide and carbonyl groups, thus effectively breaking the H-bonds, and the provision of extra space around the side chains (indicated by the specimen swelling) thus allowing more freedom of rotation about the bonds both in the main peptide chain and in the side chains. Studies with collagen and polyamides (Hiltner *et al.* 1973) showed that water can be correlated with particular molecular relaxation processes so these ideas seem reasonable. In each instance the water is plasticizing the material in the same way that plasticizers soften artificial plastics such as PVC. It is interesting that somewhat similar results can be obtained with the interaction of water with nylon, an artificial polyamide (Bretz *et al.* 1979), once again confirming the reasonableness of the basic approach to biological materials from a physico-chemical standpoint. Another interesting physical characteristic of skin keratin is its piezoelectric response when dry (de Rossi *et al.* 1986). This has led to such developments as the computer glove which records movements of the hand and thus makes possible an 'intuitive' (the nicest thing anyone can say about a computer program!) computer interface (de Rossi 1989).

Compared with the number of studies of mammalian keratins, there have been relatively few on the keratins of birds and reptiles. Those which have been made are mostly confined to feather and its constituent proteins. In terms of amino acid composition, sequence, molecular size and structure, feather keratins show very little resemblance to mammalian keratins. Astbury and his co-workers (Astbury & Woods 1933) in Leeds in the 1930s were the first to probe the molecular structure of feather keratins using X-ray diffraction. They concluded that the conformation was basically a slightly contracted form of β-sheet with an axial rise per residue of about 0.31 nm. The keratin could be strained by a maximum of about 0.06 giving an increase in the axial rise to 0.33 nm, close to the value found in α-keratin from stretched hair. The most credible model for the structure was developed in the early 1970s by Fraser and his colleagues using infra-red spectroscopy which showed that about 30% of the protein is arranged in a β-sheet with a slow helical twist (Fig. 2.14; Fraser *et al.* 1971).

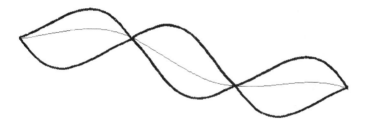

Figure 2.14 Twisted β structure of feather keratin (Fraser *et al.* 1971).

In duck and chicken feathers all the proteins have much the same molecular weight (about 11 000), much the same amino acid composition (Table 2.3; O'Donnell & Inglis 1974) and, when cross-reacted with antisera made to whole feather, show only a single immunogen. Seagull and emu (O'Donnell & Inglis 1974) and chick (Arai *et al.* 1983) feather keratins have been largely sequenced and show remarkable similarities (Gregg & Rogers 1984). Towards either end of the chain there are non-crystalline sections rich in cystine. At the centre of the chain is a large crystalline region of 62 residues which contains most of the hydrophobic and serine and glycine residues. These two amino acids have small side chains which will pack closely and they are typical of the β-sheet structures found in silks (hence serine – from the Greek for silk, *serikos*). The stability of their interactions will be increased in an aqueous environment which will tend to cause the hydrophobic structures to aggregate. The central crystalline portion is relatively conservative in amino acid composition and sequence,

Table 2.3 Amino acid composition of feather keratin

Arg	4.1	Gly	16.4
Cys	7.2	Ala	7.4
Asp	5.1	Val	8.7
Thr	4.8	Leu	2.4
Ser	10.8	Ileu	7.7
Glu	7.8	Tyr	3.0
Pro	11.1	Phe	3.7

From O'Donnell & Inglis (1974)

whereas the ends of the molecules are much more heterogeneous and variable. It thus seems that the fibre structure is strongly selected for, whereas the matrix structure can be less specific. From all this it seems likely that each protein contributes both to the structure (30% of its length) and to the matrix (70% of its length). This is therefore yet another example of a fringed micelle system. It is also very much like certain modern composites called 'block co-polymers'. In these material monomers of two types are polymerized into a single chain (e.g. AAAABBBBBBBAAAABBBBBBAAAA). When the polymer is allowed to crystallize (by a change in temperature or removal of solvent) the AAA and BBB lengths will, if properly chosen in the first place, segregate into different phases. Thus the AAA units may aggregate into fibrous phase and the BBB units go into an irregular matrix phase. Feather keratin is rather obviously doing something like this, and has been shown to crystallize ('self-assemble') *in vitro* from extracted material in solution, forming structures indistinguishable from the natural ones (Brush 1983). Other structural proteins can also be, very often, considered as block co-polymers although the differences between the zones formed may well be far more subtle than in an artificial block co-polymer. An example might be the creation of polar and non-polar zones. Obviously since amino acids can be divided into groups in so many different ways there will be many more different 'phases' into which a protein can form itself, and these considerations lead back to the acquisition of specific conformation discussed earlier.

The mechanics of feather keratin have not been much investigated. At 5 GPa (Fraser & MacRae 1980), or perhaps even 8–10 GPa (Purslow & Vincent 1978), feather keratin is at least twice as stiff as hair. This is reasonable since stresses are thrown directly on to the covalent bonds of the extended structure, whereas in hair the bonds which take the initial load are the H-bonds which stabilize the helix. These are probably fewer and certainly weaker than the covalent bonds of feather keratin in which the hydrogen bonds are aligned at right angles to the principal stresses and so only take shear loads. Feather keratin also shows much less hysteresis than hair keratin and does not show any phase changes as implied by abrupt changes in modulus. This is certainly necessary for feather rachis, which must be able to provide a wing structure which can transfer the maximum amount of the bird's energy into displacement of air. Feather rachis when tested in three point bending (a sort of beam test) shows a Hookean response with no hysteresis, right up to the sort of deflections which occur in flight. The elasticity of feather rachis is important in flight where the strain energy in the rachis keeps the wing tip moving at the end of the stroke while the wing root has started on its recovery. If you have rowed in a racing shell you will know that it is easier to finish a stroke cleanly if you have a whippy blade. This is because you can hold the handle stationary at the end of the stroke and give yourself more time in which to start the recovery whilst the whip (strain energy) in the oar is still carrying the blade through the water. The action of the feather rachis at the end of the wing stroke seems to be exactly analagous.

2.3.2 Silks

Feather keratin is primarily a twisted structure. Most silks seem also to be structures, but they are more planar and well extended (Dickerson & Geis 1969). Other conformations also occur in silks, notably some related to the α-helix and also collagen proteins (see below, Denny 1980). The amino acid composition of a 'typical' silk – that of *Bombyx mori*, the silk-worm moth – is shown in Table 2.4.

Table 2.4 Amino acid composition (%) of *Bombyx* silk

Gly	44.5	Lys	0.3
Ala	29.3	Arg	0.5
Val	2.2	His	0.2
Leu	0.5	Tyr	5.2
Ile	0.7	Phe	0.6
Ser	12.1	Pro	0.3
Thr	0.9	Try	0.2
Asp	0.3	Met	0.1
Glu	1.0		

From Lucas & Rudall (1968)

The silk protein is made up primarily of glycine, alanine and serine with a small amount of other amino acids, mostly with bulky side chains. The amount of glycine is about the same as that of alanine and serine together. This suggests the possibility of a repeating structure of gly–ser or gly–ala, with twice the amount of gly–ala as of gly–ser. Also note that both glycine and alanine are not very polar and that serine is only a little more polar. The next expectation is for a repeating structure which will be stabilized by hydrophobic interactions. X-ray crystallography allied with model building and the investigation of model materials [notably poly-L-(ala-gly) and poly-L-(ala-gly-ala-gly-ser-gly)] showed that the most probable structure is indeed the latter one. Polypeptide chains of this sequence will form anti-parallel β-sheets which, with a slight pleat in them, are stabilized by H-bonds which are unstrained (the parallel β-sheet

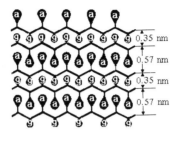

Figure 2.15 The packing of anti-parallel β-sheets in silk. Small groups marked 'g' are glycine; large groups marked 'a' are alanine.

conformation requires the H-bonds to be strained and so it is energetically less favourable and more unstable). This is shown in Fig. 2.15. The significance of the alternating residues becomes apparent when several such chains are packed together to form a fibre. The glycine side chains (– H) all project to one side of the pleated sheet, the serine and alanine chains project on the other side (a on the figure). When the chains are packed, these groups alternate as shown to give a very compact structure which is stabilized by van der Waals forces. This is summarized in Fig. 2.16 which shows the major stabilizing forces within the β-sheet areas as vectors (i.e. line length proportional to bond strength).

Figure 2.16 The major forces stabilizing silk β-sheet structure.

From this it can be seen that the fibre should be stiff along the length of the sheet, as indeed it is: about 10 GPa (Gosline *et al.* 1986) which is very similar to feather keratin. As with feather keratin the H-bonds are normal to the direction in which the principal stresses will act, so stabilizing the sheet. The van der Waals forces are relatively weak, so there is a possibility of movement between the sheets, resulting in a flexible fibre. The remaining amino acids are relatively bulky and will tend to form disordered regions between the crystalline regions. Figure 2.17 correlates extensibility and the percentage of amino acids with bulky side-groups. These disordered regions are responsible for the extensibility which silk fibres have. *Antherea* silk has sufficient amorphous material to show a hair-like stress-strain curve: a yield at a strain of 0.05 followed by a low modulus region as the amorphous areas become orientated and a final high modulus region, just before break at a strain of 0.35, showing that the

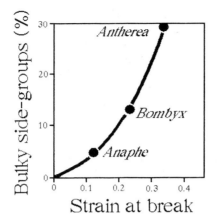

Figure 2.17 Correlation between extensibility and the content of bulky side-groups in silkworm silk.

chains in the amorphous region have been orientated by the strain. Dragline silk of the spider *Araneus diadematus* shortens dramatically when wetted and shows evidence of being rubbery, a phenomenon which has been ascribed to increased randomness in the amorphous areas (Gosline *et al.* 1984). This interpretation has been challenged (Vollrath & Edmonds 1989) as due simply to the increased effect of surface tension of the water. Certainly it seems that at the small sizes involved the forces generated at the air–water interface can be a significant component of the total. The stiffness and extensibility of the silk which a spider uses in its web are likely therefore to be controlled by three factors: the amino acid composition which will determine the amount of crystallinity possible, the distribution and amount of water in and around the fibre and the draw ratio. The draw ratio is a measure of the amount by which the fibre is extended before the stabilizing secondary forces lock the polymer chains relative to each other. With silk this presumably occurs in the presence of water which will plasticize the material and then evaporate, leaving the silk stabilized. A high draw ratio will result in highly orientated polymer chains and a high modulus fibre; a low draw ratio will give a fibre of high amorphous content, low modulus and high extensibility. That draw ratio is probably important has been shown with *Bombyx* silk, where Young's modulus and crystallinity are inversely proportional to the diameter of the thread, and it is likely that the finer thread is produced by a higher draw ratio. Silk can be 'spun' *in vitro*, when it is found not only that a minimum shear rate is required for the silk to become crystalline but that this is helped by the presence of divalent ions such as Ca^{2+} and Mg^{2+}. The average draw ratio required for silk formation in *Bombyx* is three as the silk passes from one end of the spinneret to the other (Iizuka 1966).

Silk has great commercial importance, not only for making fine clothes but also as strong fibres for medical and military purposes. At one time spider's silk was the only material fine yet strong enough to be used to make the cross-'wires' in a gun sight. Reports of its successful production in a biotechnological ferment have been common. The latest (1988) reports its production from *E. coli*, which is extracted into lithium bromide (a breaker of H-bonds) from which the silk can be spun. The idea is then to introduce regions of greater or lesser disorder so that the modulus and extensibility can be engineered and a variety of silks produced in the same way that caterpillars and spiders do.

2.3.3 Collagens

Collagens comprise a family of closely related proteins forming characteristic fibres throughout the animal kingdom (excluding Protista) and a large body of literature exists on the chemistry, structure, biosynthesis and mechanics of collagen from various animals and in various states. Collagen is the most common fibrous protein and is the basis of many glues and of the gelatin industry.

Table 2.5   Distribution of collagen in a mouse

In a mouse, 20% of the total protein is collagen	
Of this collagen, the skin contains	40%
bone and cartilage	10–20%
tendons	25%
blood vessels	5–10%
internal organs	2–8%
muscle	1–2%

Collagen occurs in most tissues and is probably the commonest protein in the body (Table 2.5). In skin and 'basement membrane' it occurs as a reinforcing fibre. It can also function as the winding of a pressure vessel, as in nematodes, earthworms and sharks. In tendon and muscle it is concerned with transmitting tensile stresses and it is in this form, notably rat tail tendon, that it has been

Table 2.6   Amino acid composition (%) of rat tail tendon

Ala	9.9	Thr	1.9
Gly	35.1	Met	0.6
Val	2.3	Arg	4.7
Leu	2.2	His	0.3
Ile	1.3	Lys	3.6
Pro	12.3	Asp	4.7
Phe	1.4	Glu	7.4
Tyr	0.5	Hyp	9.0
Ser	2.8		

From Brown (1975)

Table 2.7 Part of a collagen sequence

gly–ala–hyp–gly–pro–hyp–gly–ala–hyp–gly–ala–hyp–gly–pro–val–gly– pro–ala–gly–lys–ser–gly–asp–arg–gly–glu–thr–gly–pro–ala–gly

From Woodhead-Galloway (1975)

most studied. The amino acid composition (Table 2.6) is so characteristic that it is almost diagnostic of collagen. In particular hydroxyproline (proline hydroxylated after its incorporation into the peptide chain) is often used as a marker amino acid and for estimating the collagen content of tissues. The most striking point is the large amount of glycine, proline and hydroxyproline, all helix breakers, totalling half the protein. Statistically it seems that every third residue will be glycine, making up units with proline and hydroxyproline of the form -(gly-X-pro)- and -(gly-X-hypro)- where X is any other amino acid. Part of the actual sequence is shown in Table 2.7. This shows that glycine does occur at every third residue, but the rest is not quite so regular. The structure which this sequence forms is well within the acceptable limits on the Ramachandran plot, and is a slow left-handed helix. Three of these helices coil around each other to give the collagen triple helix. Polymers of glycine, proline or hydroxyproline will also give a collagen helix, so for some reason it is energetically favourable for these amino acids. Having glycine at every third position is important to enable the protein chains to come close together and allow hydrogen bonding between the chains, in fact a collagen-like molecule can be constructed from any three polypeptide chains which have glycine as every third residue. However, some 'driving force' is required to push the conformation in the direction of the collagen conformation – the loss of entropy in going from an irregular conformation to a collagen triple helix would be too large in the absence of proline and hydroxyproline, which direct the conformation. There is reason to think that it is possible to have too much proline and hydroxyproline, since a collagen which was too stable might be difficult to degrade during development and the dissolution which precedes healing. To some extent collagen can be compared with silk and feather keratin in that covalent bonds take the main stresses and hydrogen bonds have only to cope with shear. In doing this they are fairly labile. In order to form a fibre, the collagen molecules have to form a continuous structure with plenty of overlap to allow stresses to be passed from one molecule to the next. Just as with keratins, the basic microfibril (in this instance, tropocollagen) is assembled into larger and larger units giving a hierarchy of structure which can finally form such components as tendon (Fig. 2.18). This structure is very regular; for instance the banding pattern seen in the electron microscope and in X-ray diffraction patterns is characteristic of collagen. This banding pattern has a periodicity of 67 nm or so (depending on the source of the collagen); the individual collagen molecules are 280 nm long, and pack together in the now familiar quarter stagger pattern. This pattern can be formed *in vitro* if the collagen is precipitated from

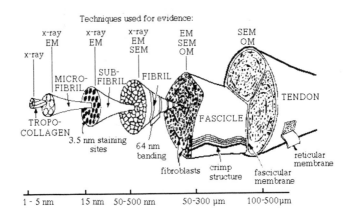

Figure 2.18 Hierarchical structure of collagen in tendon, showing the techniques used to obtain the data (Kastelic *et al.* 1978).

a weakly acid solution with 1% NaCl. The pattern is therefore determined by the amino acid sequence. This was confirmed by a computer experiment in which two collagen amino acid sequences were stepped past each other and the total hydrophobic and charge interactions estimated between each pair of amino acids. Maxima in the strength of these interactions occur at intervals corresponding to the quarter stagger (Fig. 2.19). In this sort of arrangement the collagen fibril can be considered to be crystalline, since it shows a well-defined melting temperature (about 60°C) at which the fibre shrinks to about one third its length and becomes rubbery. This is thermodynamically a first order transition and is characteristic of melting in crystalline materials.

In tendon and other structures the collagen is associated with varying amounts

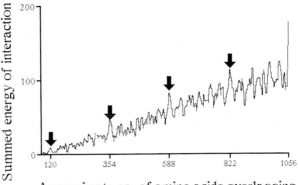

Figure 2.19 A computer plot of the number of hydrophobic and charged interactions between two collagen molecules as a function of the degree of overlap between them (Woodhead-Galloway 1975).

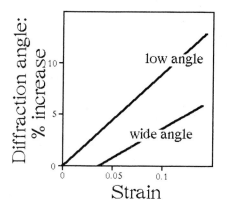

Figure 2.20 Correlation of macro- and micro-strain in stretched collagen from rat tail tendon. Wide-angle data give 0.291 nm spacing; low angle data give 67 nm spacing (Cowan *et al.*1958).

of other proteins and acid mucopolysaccharides (see Chapter 4). The matrix is concerned with the transmission of shear within the tendon and hence affects properties such as toughness. However, the tensile properties of tendon can be accounted for mostly in terms of the collagen triple helix which comprises 70–80% of the dry weight of the tendon. This has been investigated by X-ray diffraction of stretched samples (Fig. 2.20). The small angle diffraction pattern, corresponding to the 67 nm banding, increases linearly with strain, accounting for 90% of the strain up to strains of about 0.15. However, the wide angle spacing, corresponding to the atomic spacing within the protein chains, does not increase nearly so quickly with strain, indicating that although the helix itself is being stretched, this can account for only about half the total strain. The conclusion from these experiments is that the tendon is a two (at least) phase material with relatively low extensibility in the crystalline regions and relatively high extensibility in some less organized regions. Presumably the less organized regions are the matrix component. Removal of the polysaccharide component of the matrix makes little difference to the mechanics of the tendon; removal of the non-collagenous protein fraction reduces both the initial modulus and the ultimate strain by a factor of ten.

One further point requires consideration before looking at the mechanical properties of bulk collagen in a tendon. That is the effect of water. Dry collagen is brittle and stiff with a Young's modulus of about 6 GPa. This is stiffer than hair and about as stiff as feather keratin. It is not as stiff as silk. Addition of water softens the collagen progressively. This interaction of water has been shown to occur in a well defined manner. The most strongly interacting water is probably incorporated as an integral part of the structure of the triple helix, probably about two water molecules for each tripeptide. When the water content exceeds two molecules per tripeptide the molecule starts to swell laterally.

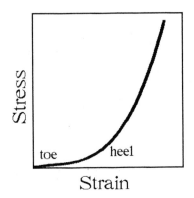

Figure 2.21 Typical stress–strain curve for a collagen-containing biomaterial.

At this level of hydration the water is having a marked plasticizing effect and is present in about 25% w/w. It is probably associated with the charged side chains. At greater water content, more is associated with the matrix material. The behaviour of water in association with collagen is in some ways similar to its behaviour in association with nylon and stratum corneum.

The stress–strain curve of materials containing collagen, such as a tendon, is very typical (Fig. 2.21). There is a toe region of low modulus which curves round to give a region of much higher modulus. This latter is the normally quoted modulus. The toe part is normally attributed to the unfolding of crimp structures (Fig. 2.18), and this interpretation has been well confirmed with both polarized light studies and with mathematical analyses (but see Section 4.4.1). The length of the toe region varies with the source of the tendon and therefore, presumably, the amount of crimping varies in different collagens. The modulus of a wet tendon in the steep part of the curve is typically 1.5 GPa (Bennett *et al.* 1986).

Collagen in tendon is used in two main ways – at high and medium strain rates and with static loads. In both these modes of use it is unlikely that the collagen is ever totally unloaded. It is much more likely that even in the tendons of a running animal the strains are such that the material is spending most of its time working in the high modulus region, even when the associated muscles are not contracting. This is even more obviously true for collagen in arteries which are, as simple dissection will show, in a state of prestress. Apart from maintaining the collagen in its high modulus region, the importance of such prestress is obscure, even though it seems to occur in many tissues other than arteries. At the high strain rates which will occur during running and walking it is likely that the modulus will be higher and that viscous loss processes will be less important as there will simply not be enough time for them to occur. It is also probable that the toe region will be less evident. At lower strain rates with a constant rate of extension, conditions under which most tests are performed on collagenous tissues, the modulus is much lower, and therefore

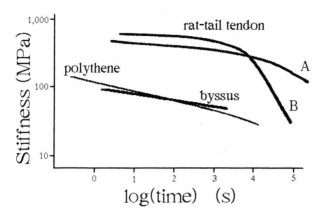

Figure 2.22 Stress-relaxation curves for two types of collagen (thick lines) and polyethylene (thin line). A, 0.035 strain; B, 0.075 strain (Wainwright *et al.* 1975, Smeathers & Vincent 1979).

extensibility up to the ultimate stress will be greater. It is therefore probably dangerous to extrapolate results from medium strain rate experiments to situations in which the strain rates are much higher. Under sinusoidal straining at rates varying from 0.2 to 11 Hz there is little or no dependence of modulus or energy dissipation on frequency (Bennet *et al.* 1986). The whole question of how collagen functions *in vivo* is a very vexed one. Tendon is also used to withstand static loads as in the Achilles tendon of a standing cow or man. Stress--relaxation tests performed on rat tail tendon (Fig. 2.22) show that the tendon does stress–relax, but that within the limits of reversible stress–strain behaviour, tendon performs very similarly to bulk crystallized polyethylene. This confirms the crystalline nature of collagen, since such a broad spectrum of relaxation times is typical (though not necessarily diagnostic) of a crystalline material. At a strain of 0.075 the tensile behaviour is no longer reversible, resulting in breakdown of the tendon at long times. It might be thought that since rat tail tendon is not normally loaded for long times statically, that Achilles tendon and other such collagenous structures would be more tightly cross-linked so

Table 2.8 Amino acid composition (%) of *Mytilus* byssus thread

Ala	12.9	Met	0.9
Gly	29.5	Arg	4.0
Val	3.8	His	2.1
Leu	3.6	Lys	4.0
Ile	2.1	Aps	6.8
Pro	6.4	Glu	6.1
Tyr	1.3	Hyp	3.1
Ser	6.6	Cys½	1.1
Thr	3.9		

From Pujol *et al.* (1970)

that they could achieve an equilibrium modulus of useful value. A useful material for comparison is the byssus thread of *Mytilus*, the common mussel. Fifty to a hundred of these threads guy the mussel down, resisting wave action, prestrained by about 0.1 (Smeathers & Vincent 1979). They are thus continually under tension. X-ray diffraction, infra-red spectroscopy and amino acid analysis (Table 2.8) show that the thread contains significant amounts of collagenous protein. The sum of glycine, proline and hydroxyproline is 39%. Thus it is possible that about 55% of the thread is collagen. Some of the remainder may be β-protein. The stress–relaxation behaviour, shown in Fig. 2.22, is very similar to polyethylene. The material is highly insoluble, requiring hot concentrated acids and alkalies to take it into solution. It is stabilized with phenols, but this seems not to reduce its rate of relaxation much compared with the hydrogen-bonded collagen. Yet this thread is continually in tension. From this it might be surmised that the collagen of Achilles tendon need not be much more cross-linked or otherwise stabilized compared with rat tail tendon. It may be that our concept of 'static load' needs to be revised and that no such thing really exists in nature – nothing is still enough for long enough. The stress–relaxation test would then become inappropriate as a model for the working conditions of the tendon or byssus thread. The collagen of *Ascaris* is possibly one of the few which is really cross-linked. It contains 2.7% cystine residues which represents a potentially very high cross-linking density. This would be required since nematodes rely on hydrostatic pressure for all their locomotive functions and so really are continually stressed.

There are several artificial systems in which collagenous materials are highly stressed continually: gut violin strings, calf skin banjo vellums, bow strings and the like. The strings of musical instruments relax for a few hours, then reach an equilibrium and are more at the mercy of moisture, grease and changes in temperature for their changes in tension. Drum and banjo vellums are taken to much higher stresses and relax considerably for the first week. After that they may require monthly attention. They are not much different from plastic vellums at this time, which can relax quite a surprising amount over the years. Keeping to the musical theme it has been pointed out that the collagen of gut strings has a low hysteresis and therefore absorbs very little of the energy used to sound it and will vibrate for some time. However, the hair of a violin bow has high hysteresis and so can absorb the energy imparted to it and not resonate. Consequently one hears only the violin and not the bow.

2.3.4 Protein rubbers

The proteins discussed so far have been fibrous with a regular periodicity along the peptide chain which will form regular structures. The major bonds within these fibres are orientated along the main axis of the fibre: consequently these materials have a relatively high modulus and low extensibility, typical of a crystalline material. What strain they can accommodate in excess of (say) 0.05 is due to the presence of disorder or amorphous regions. However, there are some proteins which are more extensible and have a lower modulus much more like that of a rubber. The most studied are resilin, abductin and elastin. The mechanisms of elasticity of these 'rubbers' seem, however, to be different.

2.3.4.1 Elastin

Elastin is the main elastic protein of vertebrates and is usually found in association with collagen. It is very stable when heated and is almost defined as that fraction remaining when collagen and other components have been removed by autoclave treatment at 110°C. It is most commonly prepared from the yellow neck ligament (ligamentum nuchae) of ungulates, which counterbalances the weight of the head in the same way that some 'up-and-over' garage doors are counterbalanced by large springs. This considerably reduces the muscular effort required to move the head. The ligamentum nuchae is about 80% elastin, has a shear modulus of around 0.6 MPa and a stress–strain curve shown in Fig. 2.23 (Dimery *et al.* 1985). The amino acid composition of elastin is shown in Table 2.9. The last three amino acids listed are covalent cross-linkers. As a result of this cross-linking, elastin is rendered largely insoluble and can sustain loads. As with the other proteins discussed, water is essential to elastin as a plasticizer.

There has been considerable argument as to the nature and origin of the elasticity of elastin (Gosline 1980, Urry 1983). Two of the rival theories, now apparently defunct, required elastin to have a random structure, an attribute which is supported by two lines of evidence: nuclear magnetic resonance studies show that the chains are kinetically free and supposedly therefore not involved in any fixed structures. Confirmation of the lack of fixed structure was given by X-ray diffraction, which shows no spots which would indicate short-range structure and which also shows that on extension the protein does not become

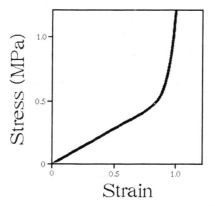

Figure 2.23 Stress–strain curve for ligamentum nuchae (Dimery *et al.* 1985).

Table 2.9 Amino acid composition (%) of elastin

Asp	0.7	Tyr	0.6
Thr	1.0	Leu	6.5
Glu	1.7	Phe	3.4
Pro	12.5	Lys	0.4
Gly	31.6	His	0.05
Ala	21.3	Arg	0.7
Cys½	0.4	Hyp	0.7
Val	13.4	Isodesmosine	0.2
Met	2.7	Desmosine	0.1
		Lysinonorleucine	0.1

From Franzblau (1971)

appreciably orientated or crystalline, but remains essentially random. These observations, together with low modulus and high extensibility, suggested that elastin is wholly, or nearly, rubbery, and that its mechanics can be explained quite adequately by rubber elasticity theory (Hoeve & Flory 1974). However, the amino acids of elastin are largely (60%) non-polar and only 5% are polar. This led to the proposal that in an aqueous environment the conformation was globular, rather like droplets of oil in suspension (Weis-Fogh & Andersen 1970). This interpretation was supported by several observations: in less polar solvents, such as mixtures of formamide and water, the stiffness drops as the stabilizing hydrophobic forces in the 'oily' centre of the globule are disrupted and the protein chains gain greater freedom of movement. Also a fluorescent probe – a molecule which changes its fluorescence according to the polarity of the surrounding medium – can be bound into the hydrophobic region of the elastin molecule. Changes in its fluorescence as the elastin is extended show that extension is accompanied by an increase in hydration of the hydrophobic region, a conclusion also supported by thermodynamic data (Gosline 1980). However, sections of repeating sequences of peptides – the pentamer (Val-Pro-Gly-Val-Gly) and the hexamer (Ala-Pro-Gly-Val-Gly-Val) – and the observation in the electron microscope of fibres in preparations of dispersed elastin, suggested that randomness was not pervasive (Urry 1983).

Some philosophy: one of the basic tenets of molecular biology (or, for that matter, biology as a whole) has been that where anything exists it exists for some reason. It is highly unlikely that evolution tolerates junk. An example of this is any enzyme. Naively considered as a functional unit, the only bit needed is the active site. The remainder of the protein appears to have no function until it is realised that the environment and support of the active site have their own importance. These are provided by the rest of the enzyme molecule. The regularity of primary structure of fibrous proteins has been shown to have significance in determining secondary and tertiary structure and hence the mechanical properties of the material. Elastin shows patterns of amino acid sequences. Current dogma says that it is likely that this pattern implies structure,

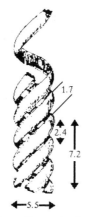

Figure 2.24 Overall dimensions (in nm) of a fibre of elastin (Urry 1983).

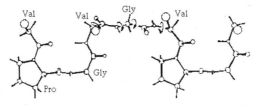

Figure 2.25 β-turn of elastin, showing the rotations which lead to the elastic properties (Urry 1983).

Figure 2.26 Assembly of β-turns into the primary elastin helix which is further assembled into fibres as in Fig. 2.24 (Urry 1983).

which in turn makes it more unlikely that elastin is totally rubbery and random. This was the position adopted by Urry, whose model for elastin can accommodate both experimental observation and philosophy.

His main problem was to reconcile the apparently strong patterns indicated by the presence of fibres and molecular structure with the data suggesting rubbery elastic mechanisms. He did this by proposing a coiled coil structure (Fig. 2.24). Such a structure will form fibres, yet will not give a well-defined X-ray diffraction spectrum because the structure is so open and 'long range'. The specific amino acid sequences give rise to 'β-turns' (Fig. 2.25) which can be strung together to give a helical structure sufficiently open to allow water to penetrate the core (Fig. 2.26) and which forms each of the three units of the coiled coil (Fig. 2.24). The reality of this hypothesis is supported by a large amount of chemical and physical analysis and modelling. The elastic properties reside primarily in the rotational freedom of the bonds in the β-turn (indicated in Fig. 2.25), the so-called 'librational entropy mechanism'. This freedom of rotation is assured by the hydrophobic nature of the protein and the large amount of water entrained. It also explains why elastin was originally thought to be rubbery: the β-turns are so flexible that they can adopt a wide variety of shapes and so can be modelled as entropic – which to a large extent they are, but not in the sense of rubber elasticity.

2.3.4.2 Resilin and abductin

Resilin and abductin are dealt with together since the available evidence suggests that they both conform quite closely to classical rubber elasticity theory (Section 1.2), despite the fact that they need to be plasticized with water for them to work at normal temperatures and that this plasticization swells the material, thus straining the 'random' chains. Abductin occurs as the inner hinge ligament of bivalves, where it acts as the (passive) antagonist to the shell adductor muscles (Kahler *et al.* 1976). It functions as a simple rubber pad in compression. Resilin occurs in insects and possibly other arthropods where it serves as a store for strain energy (Jensen & Weis-Fogh 1962). For instance, in the flight mechanism of the locust, resilin in the wing hinge is deflected at the extremes of the wing strokes, slowing the wing down and storing the kinetic energy of the wings are strain energy. The stored energy is then delivered back to the wing for the start of the next stroke where, as kinetic energy, it enables the wing to accelerate more quickly into the next stroke. Resilin is also used to store energy delivered to it at a relatively slow rate by muscular contraction, delivering it at a far faster rate, thus enabling the flea (amongst other insects) to jump (Bennet-Clark & Lucey 1967). Amino acid compositions of resilin and abductin are shown in Table 2.10. All of these have a large percentage of helix-breaking residues and indeed thermodynamic experiments strongly suggest that the protein chains in these two materials are in totally irregular conformation (Section 1.2).

Both resilin and abductin act as mechanical energy stores. Both are able to recover completely after applied deformation, implying that the cross-links (tyrosine derivatives in both materials) are covalent (Andersen & Weis-Fogh 1964, Andersen 1971). The covalent cross-linking also accounts for the insolubility of the proteins in anything which does not break covalent bonds, a fact which has prevented studies of the amino acid sequence of these materials.

Table 2.10 Amino acid composition (%) of two invertebrate protein rubbers

	Abductin			Resilin
	Placopecten	Ensis	*Mytilus*	Locusta
Asp	1.7	23.9	3.9	10.8
Thr	1.1	1.6	1.4	3.0
Ser	6.0	1.9	8.9	8.1
Glu	1.3	3.4	3.7	4.8
Pro	1.1	7.2	4.7	7.9
Gly	68.4	29.5	41.1	36.7
Ala	3.0	4.6	4.2	10.6
Val	0.5	3.0	2.2	3.2
Met	6.0	13.5	9.7	—
Cys½	0.2	1.5	1.3	—
Ile	0.6	0.8	3.1	1.7
Leu	0.2	3.5	7.2	2.4
Try	0.1	—	0.4	3.1
Phe	8.6	2.6	2.2	2.4
Lys	0.9	2.0	3.4	—
Arg	0.4	0.6	2.2	—
His	—	0.6	2.2	—
Helix breakers	87.2	61.5	58.6	63.5

From Kahler *et al.* (1976), Andersen (1971)

However, not all the energy put into deformation (measured as the area under the stress–strain curve on extension) is recoverable (measured as the area under the stress–strain curve on retraction to zero stress and strain). The ratio of energy recovered to energy put in is the resilience, R

$$R = 1 - 2h$$

where $h = \pi \tan\delta$, and is the elastic loss factor. The function $2h$ is also equal to the area of the hysteresis loop which is analagous to the stress–strain loop derived in dynamic tests (but note that the hysteresis loop is obtained from a straight line, or ramp, function. Dynamic tests are conducted with sinusoidal straining, Section 1.4.3). R, then, is a measure of the efficiency of the rubber. Resilin has a high resilience of 96–97%, which is as good as or better than the very best synthetic rubbers (Gosline 1980). It is important that this should be so, because energy dissipated will appear mostly as heat. So the perfection of resilin is important not only to increase the efficiency of the flying animal but also to stop the wing hinges overheating.

The variety of abductins is much greater. Swimming bivalves which open and close the shell several times a second require an efficient hinge ligament and the resilience is as good as that of resilin. Sessile and digging bivalves do not move their shells so often or so fast and this is correlated with a lower resilience, nearer 80% (Kahler *et al.* 1976). This is probably at least partly related to the amount of glycine present (Table 2.10).

2.4 POINTS TO PONDER

It may be taken as one of the basic tenets of the study of biomaterials that no mechanism be proposed for a biomaterial which cannot be shown to exist in other non-living systems: vitalism is never a successful theory (Gordon 1980). In some ways this is a rather difficult tenet to maintain since biological materials are probably much more complex than artificial materials. However, it is always comforting to find that a mechanism proposed for a specific instance is not unique, since it makes the idea of the mechanism that much more credible. Urry has produced artificial model elastins.

For instance a problem which has never been satisfactorily explained is how elastin, resilin and abductin manage to exhibit such perfect elasticity. Undoubtedly the chemistry of the protein has a large part to play: for instance the amino acid composition of resilin and abductin suggests a high degree of random conformation. This is confirmed by thermomechanical tests and (for resilin) by X-ray diffraction. Does this necessarily and uniquely imply the total randomness which works so well as a model for *Hevea* rubber? The side chains of the amino acids in resilin are mostly rather bulky and must present many potential sites for the formation of hydrogen bonds. The reason why H-bonds appear not to exert an effect has not been discovered; they are presumably all masked by the presence of water as a plasticizer. Since the primary structure of these proteins is constant, or probably so for all the constituent precursor molecules, then the cross-link sites are rigidly defined, and may well be spaced very evenly. This is never so for the rubbers of commerce in which the molecular weight between cross-links is defined by some Gaussian function. The implications of control of cross-link sites have not been explored in classical rubber elasticity theory. No bits of resilin or abductin have yet been sequenced, but if any pattern emerges in the primary structure the dogma of molecular biology would consider it to cast doubt on the rubbery nature of these proteins. How do you detect random primary structure? Is the belief that all primary pattern leads to stable defined structures merely a belief, despite the success achieved by its application? And as the lessons of elastin show, it is not necessary to have random structure in order to invoke rubbery mechanisms of elasticity. My suspicion is that resilin and abductin will turn out to be as ordered as elastin, despite the ease with which they can be described by a model based on statistical disorder. At the very least, resilin must have an area of β-sheet structure with which it can interact with and stabilize the chitin usually found associated with it. This will be of significance to the developers of materials in robotics, who are currently (1990) looking at protein rubbers as inspiration for new types of 'perfect' rubber for use in various sensory and suspension systems.

Perhaps the latter people should take note that protein rubbers are laid down in the presence of a swelling agent – water – which is probably present in much larger amounts during synthesis and cross-linking than in the functioning material. In artificial systems a swelling agent has been shown to reduce the deviation of the elastomeric properties from the theoretical predictions (Section 1.2). Probably the resulting elastomer has fewer random imperfections such as trapped entanglements. However, the primary sequence of the protein itself presumably contributes to the lack of random imperfections.

2.5   WHAT HAPPENS TO ALL THAT STRAIN ENERGY?

Collagen, elastin, resilin and abductin all function as stores of strain energy for use during locomotion and are by no means the only materials to do so. These materials, to be of any use as an energy store, have to be resilient. But this will make the material brittle since if the material has no way of absorbing energy then that energy can be used in crack propagation and the creation of new surfaces (Section 1.6). Insect cuticle seems to be a good example of this type of system. For instance the apodeme of the hind leg of the locust forms a major part of the energy storage system for the jump of the locust (Bennet-Clark 1975). Because the tendon is so stiff (Section 5.2) it stores energy at relatively low strains – about 0.03 – but can store twice the amount of energy that resilin can since it is 200 times stronger than resilin and works at 1% of the strain at which resilin works. The tendon is loaded to within 15% of its ultimate strength before a jump but this is quite safe (mostly!) since the tendon is loaded relatively slowly. At a rough estimate the size of crack which a piece of resilin could accommodate without fracturing under load would be, at 0.25 mm, rather larger than that at which a locust apodeme could accommodate under comparable conditions (i.e. 0.1 mm) for a similar amount of energy stored. These measurements are not necessarily smaller than the size of the pieces of cuticle in question, so resilin pads are at risk from catastrophic failure only in larger insects. This is not too bad if they are kept out of the way as, say, is the tendon down the middle of the hind leg of the locust. Even so it is difficult to load the apodeme experimentally to anything like its natural strength and this has been attributed to surface cracks initiated during the preparation of the specimen (Bennet-Clark 1975).

By contrast, the collagenous tendon of vertebrates is about a sixth as strong as locust tendon and a tenth as stiff. But because it can work at higher strains it can also store large amounts of strain energy. Both collagenous and cuticular tendons have to work at low strains since they are in series with muscles whose efficiency will drop off if they have to contract too far: the locust tendon probably needs to be stiffer because the muscle is more rigidly contained and so will not be able to change shape to any great degree. Collagenous tendons differ in another way from the other energy stores so far mentioned: they are fibrillar. This is important since although the tendon can have resilience of 90–94% (Bennet *et al.* 1986). At strains above 0.04 the tendon can absorb energy by the lateral separation of fibres (defibrillation) within the bundles.

The resulting damage can be assessed and related quantitatively to the amount of energy absorbed. In addition, if the matrix is relatively compliant it will not transfer shear stress so readily from one fibre bundle to the next so that the normal mechanisms of crack propagation will be frustrated [much the same sort of thing happens in grass leaves (Section 5.2.2) and in rope]. So the tendon can absorb energy in a manner which does not affect the longitudinal continuity of the fibres and hence the load-bearing capacity of the tendon. At higher loads, the fibres start to break and pull out, absorbing even larger amounts of energy. Thus the fibrillar architecture of tendon gives a highly resilient yet very tough energy store – a great advance over resilin though not so dissimilar from locust leg tendon.

One can see why the difference between these energy storage materials can exist when one considers their morphology in use. Resilin and abductin are nearly always used in compression or bending and are rarely loaded in tension. Thus the low toughness is not a disadvantage because the material is rarely exposed to conditions favourable to crack propagation. But in the dragonfly there is a pure resilin tendon which is loaded in tension as part of the flight system (Jensen & Weis-Fogh 1962). Its margin of safety has not been calculated although, like the locust apodeme, it is working within the insect and so is unlikely to be exposed to scratches which might initiate cracks. The other unusual thing about this way of using resilin is that it is a high-strain rubber being used in series with a muscle – superficially a 'bad thing' (see above). Presumably the wing muscle of the dragonfly is attuned to the resilin in some way, but this has not been investigated.

In contrast to most resilin and adbuctin systems, the leg tendons of a running sheep or man are being loaded in tension and at a high strain rate, possibly even equivalent to impact loading in a sprinter. Under these circumstances the possibility of overloading is much greater than in the locust where the tendon is loaded in a relatively slow and controlled manner, although such tendons are rather subtly matched to their working loads and strains (Ker *et al.* 1988). The human Achilles tendon can be fractured if the ball of the foot is loaded very quickly as in a sprint start off blocks, or a stumble down stairs. A sub-critical load can cause excruciating pain. As a corollary the abductins of non-swimming bivalves are much less resilient, which may indicate that the hinge ligament is being selected for toughness in holding the two shells together.

The resilience of tendon seems to be independent of temperature between 20 and 36°C (the range measured so far). Elastin, which is as resilient as resilin, also retains its resilience over a wide range of temperature (Fig. 2.27, Gosline & French 1979). If it is kept at a constant level of hydration during testing by immersing it, hydrated, in liquid paraffin rather than saline, it shows a glass transition (Section 1.5) the data being obtained at different temperatures and superimposed according to the WLF time–temperature superposition formula. But if the elastin is tested in an aqueous environment it shows no glass transition since, because of its hydrophobic nature, it swells as the temperature

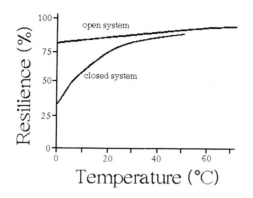

Figure 2.27 The resilience of elastin varies with temperature (closed system), but if water is allowed in freely (open system) it acts as a plasticizer and depresses the glass transition temperature: thus the elastin does not change its modulus so much (Gosline & French 1979).

falls (Fig. 2.28). Water is a plasticizer, effectively lowering the glass transition temperature (T_g) in proportion to the amount of water present. Since at T_g, tan δ is high, resilience will be low: but away from T_g, E'' will be lower, tan δ will be lower and resilience be higher. (Elastin may not be the only material to show such behaviour. Collagen has zones of hydrophobicity which serve to 'stick' the triple helices together. These areas will presumably change their reaction with water as the temperature changes.)

So the resilience of elastin is insensitive to changing temperature: this is presumably of more significance in animals which cannot regulate their temperature such as the lower vertebrates. The elastic mechanism of elastin has implications in disease: one of the factors in arterial disease is the deposition

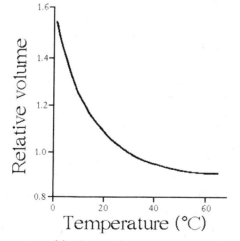

Figure 2.28 Water enters elastin as the temperature drops due to elastin's hydrophobicity (Gosline 1980).

of fatty deposits (plaques) on the artery wall. If these fats can enter the elastin molecule they could not only disrupt the normal elastic mechanism of elastin but raise the T_g and make the material less resilient. This would make the elastin more susceptible to fatigue fracture since it would be absorbing more strain energy – certainly the elastin breaks down in diseased areas and the change in mechanical properties of the artery can lead to lower fracture energy and so to greater danger of catastrophic failure of the artery wall.

CHAPTER 3

Sugars and Fillers

At the outset it should be realised that relatively speaking less is known of the polysaccharides than of the proteins. This is largely a chemical problem – amino acids have much more convenient handles for chemical processing and identification. They are also more intimately associated with the genetic code, which has persuaded biochemists and biologists that protein structure is more important and basic. Ironically it is the sheer chemical anonymity, combined with extreme abundance, of polysaccharides in general which makes them commercially far more significant – as cellulose and chitin (natural and processed), carrageenan, agar and the like. Such products are used for building, making paper, stabilizing food substances, making fibres of all kinds for clothes and ropes and in the dyeing industry. As a general reference, use Rees (1977).

3.1 SUGARS

The polysaccharides which have any mechanical significance are made of hexoses which can occur in any one of four main conformations (Fig. 3.1). With glucose (β-D-glucopyranose) the commonest or most likely conformation is the $^4C_1$ chair, since this has the lowest internal energy. Its bonds are least strained in this conformation and the various groups sticking out of the ring interfere with each other to the least extent. The sugar shown is the β-form with the OH on carbon number one above the plane of the ring. The α-form has the OH and H on this carbon reversed. The D in the name means that the molecule rotates the plane of polarisation of incident light dextrally or to the right when it is in solution. The laevorotatory or L-form of the pyranose sugars is rarer in biological materials but it does occur, for instance in the pectins which stick plant cells together, which contain L-rhamnopyranose (rhamnose). The basic sugar unit can have its chemical properties changed by changing the side-groups. These changes are reflected in the different names given to the resulting sugars. For example, if the OH on C(2) is replaced by the group $-NHCOCH_3$, the result is β-D-N-acetylglucosamine; if the H_2OH on C(6) is replaced with $-OOH$, the result is β-D-glycuronic acid; if the same group is replaced with

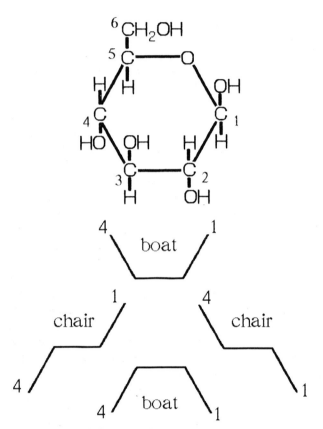

Figure 3.1 (upper) Haworth formula of hexose. (lower) Boat and chair conformations of hexose.

H_2SO_3H – the result is the acid sulphate. These and more are shown in Fig. 3.2. In this way it is possible to produce sugar units with a variety of charge, chemical and steric properties. A further variable is the way in which the sugar units are linked together. Amino acids are linked only through the peptide link and this has fairly consistent steric properties. Sugars can be bonded, by a condensation reaction, by any of the OH groups in the molecule. In pyranose there are five of these on C(1), C(2), C(3), C(4) and C(6) and each of these can exist in α or β form. So there are ten possible bonding sites and 100 ways in which two pyranose molecules can be linked to form a disaccharide. In practice this number is much less but it does suggest that there must be very strict control over the way the monomer units are assembled. The bonds are named according to whether they are α or β and their position on the sugar ring. Thus the cellobiose (biose = two sugars) molecule of Fig. 3.3 is made of two β-D-glucopyranose units joined by a β-1,4 linkage. Just as with the peptide bond there is freedom of rotation about the linkages joining the sugar rings so a conformation map can be drawn in a very similar manner to a Ramachandran plot (Section 2.2). The two angles are again called ϕ and ψ. Fig. 3.4 shows the conformation map for cellobiose. The point X corresponds to the conformation found experimentally by X-ray diffraction in crystals of β-cellobiose, C marks the cellulose conformation the T the theoretically most probable (lowest energy) conformation. There is no experimental evidence that the β-1,4 linkage ever exists in any significant amount in the smaller zone (unmarked), either in the solid state or in solution. The linkage is further stabilized by hydrogen bonding. Such bonds can occur between any O and H, so there is a far greater range of stabilized conformations than with proteins, where the H-bonding possibilities are much more limited. Fig. 3.5 shows the H-bond which normally occurs with a β-1,4 linkage. Obviously this sort of conformation will be more stable in the crystalline state: in an aqueous environment the water molecules will be competing for the H-bonding sites and the rotation about the two bonds joining the sugar rings will be much greater since the H-bonds between the rings will be less permanent. There will therefore be a tendency towards a random coil formation. However, as with proteins, there exists a number of types of conformation which are extensively stabilized by H-bonds and which have been detected in crystals and characterized by X-ray diffraction. They have been shown, in several instances, to occur in solution as well. The reason is the same as with proteins – that a regular structure such as a helix which is stabilized by H-bonds and other forces is more stable as it increases in size and cooperative interactions become more significant.

Linkages between any of the ring carbons have only two bonds about which to rotate. But linkages involving the C(6) on the side chain will obviously be much more mobile, not only because there are three bonds about which the units can rotate but because the two rings are further apart and are thus less likely to interfere with each other. In structural biopolymers this linkage is commonest at the junction of polysaccharide chains to protein chains, which seems mostly to occur at serine and asparagine residues.

The linkages will define the conformation of the polysaccharide. For a good summary of this, see Rees (1977). Table 3.1 summarizes the consequences of the commonest linkages and the conformations produced if only one type of linkage is present.

Many structural polysaccharides contain more than one type of linkage and more than one type of residue. There are two main ways in which the residues can be arranged along the polymer; periodic (ABABABA....) and interrupted (AAAAAABBBBAAAABBBBBBAA...), the former having a strong pattern in the arrangement of residues, the latter having the residues arranged in blocks of varying length. Since the types of linkages between the residues may vary it is difficult to predict the conformation of these molecules. So a combination of computer modelling and X-ray crystallography has been resorted to and found successful. These techniques have shown that periodic chains with mixed linkages can form anything from undulating ribbons to extended hollow helices.

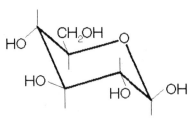
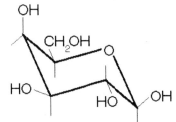

β-D-glucopyranose   β-D-galactopyranose
(glucose)   (galactose)

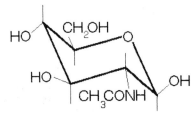
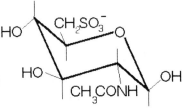

2-acetamido-2 deoxy-β-D-glucose   β-D-NAG sulphate
(*N*-acetylglucosamine)

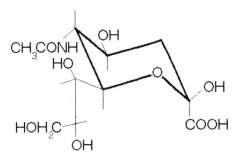

N-acetylneuraminic acid

Figure 3.2 Some hexoses found in polysaccharides occurring in biological materials, shown as conventional formulae.

In summary, a comparison of sugars with amino acids as building blocks reveals two major areas of difference.

1. There is far less variety in side chain types of polysaccharides as regards size, conformation and polarity/charge characteristics. There are no hydrophobic interactions. There is thus the possibility of high hydration and/or hydrogen bonding or ionic interactions.

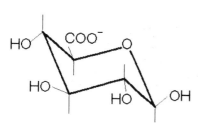

β-D-glucuronic acid

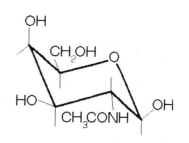

β-D-*N*-acetylglucosamine

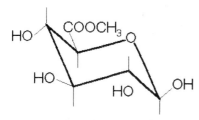

β-D-methyl glucuronate

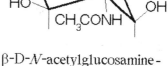

β-D-*N*-acetylglucosamine-
4-sulphate

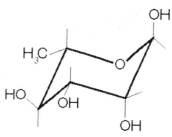

α-L-rhamnopyranose

Figure 3.2 *continued*

2. There is a far greater variety of bonding available between polysaccharide residues. This means that a greater variety of periodic structures can be generated. These structures occur over lengths of chain which are sufficiently long to allow weak attractive forces to accumulate and to have the strength of a stable and essentially permanent bond when two or more chains of complementary structure come together.

There are three main types of biomaterial which are exclusively polysaccharide. These are fibres, elastic gels and viscoelastic gels. To some extent, mainly under experimental conditions, these states are interconvertible. Thus it is possible

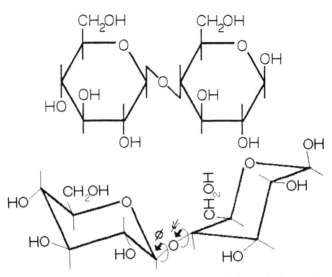

Figure 3.3 Structure of cellobiose shown as Haworth formula (upper) and showing rotations about the linkages between the sugar rings in a conformational formula (lower).

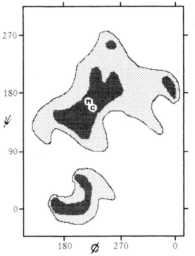

Figure 3.4 Conformational energy map for cellulose (contours correspond to 5 and 10 kcal mol^{-1}). M = lowest energy state; C = cellulose (Rees 1977).

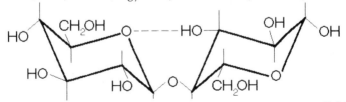

Figure 3.5 H-bonding associated with the β-1,4 linkage in cellobiose.

Table 3.1 Effects of sugar–sugar linkages on conformation

Most likely conformation	Geometrical relationship of residues	Linkage
ribbon	zig-zag	β–1,4 β–1,3
		α–1,3 α–1,4 (glucose, galactose)
hollow helix	U-turn	β–1,3 α–1,4
	twisted	β–1,2

From Rees (1977)

to form fibres from elastic gels (e.g. carrageenan) and viscoelastic gels (e.g. hyaluronic acid) and this has enabled their study by X-ray diffraction. It is also possible to produce gels from fibres – e.g. from chitin or cellulose, and this is the basis of a number of industrial processes.

3.1.1 Fibres

Of the fibrous polysaccharides, the most abundant are chitin and cellulose. These two have very similar primary structure (Fig. 3.6). There is some evidence to suggest that every sixth or seventh residue of chitin is not acetylated, but apart from that the two polymers are homogeneous. Both chitin and cellulose are probably polymerized just outside the cell membrane, possibly 'spun' off the polymerase enzyme in such a way that they can immediately polymerize into the highly hydrogen-bonded elementary fibril with H-bonds along and between the chains. As with protein β-sheets it is possible to have parallel and anti-parallel forms though the latter is not natural for cellulose. The cellulose elementary fibril is about 3.5 nm diameter containing about 40 molecules; that of chitin (in insects) is commonly about 2.8 nm containing about 20 molecules. Elementary fibrils of up to 300 nm have been reported for chitin in insects, and up to 25 nm diameter in crustacea. Cellulose elementary fibrils can be arranged into larger fibrils 20–25 nm in diameter: this might be the case in crustacean chitin.

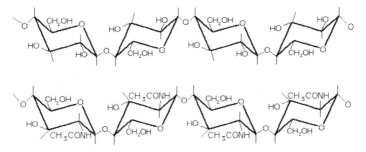

Figure 3.6 Comparison of structures of cellulose (top) and chitin (bottom).

Both polymers have a high modulus. Cellulose is the most-studied and, using X-ray diffraction to estimate strain (cf. collagen, Section 2.3.3) during tensile tests, a modulus of 140 GPa was estimated. A slightly higher figure has been obtained by calculation from the crystal, considering straightening of the covalent bonds and stretching of interchain H-bonds. If the H-bonds are ignored the calculated modulus drops by a factor of 8 or so showing that the H-bonds are making a significant contribution to the stiffness. Chitin would be expected to have a higher stiffness still, since the acetyl group not only provides from more H-bonding but reduces the flexibility of the linkage due to steric hindrance (Fig. 3.7; Minke & Blackwell 1978). The actual stiffnesses of bulk cellulose and

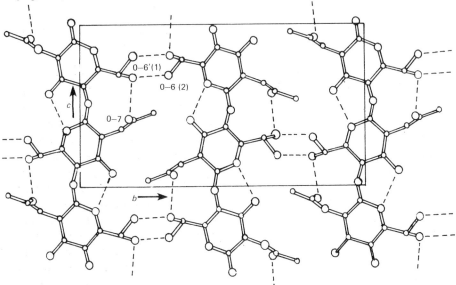

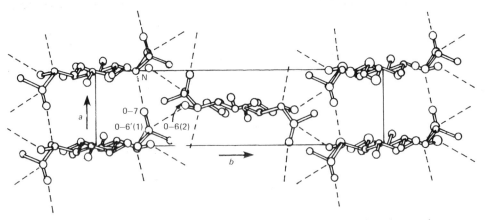

Figure 3.7 Detailed crystalline structure of α-chitin. Hydrogen bonds are shown as dotted lines (Minke & Blackwell 1978).

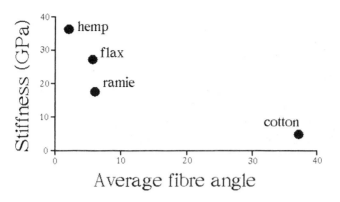

Figure 3.8 Variation of stiffness of plant fibres with the angle the cellulose fibres make with the longitudinal axis.

chitin will be lower than these theoretical and ideal values since the degree of crystallinity is rarely perfect and the orientation of the fibres will not be strictly parallel. Nevertheless, dry flax has a modulus of 100 GPa; the best estimate for the stiffness of chitin is that for the extensor apodeme of the locust hind leg which contains highly orientated chitin. Its modulus is about 80 GPa when wet (Ker 1977). This is about the same as wet flax. The stiffness of cellulose in plant fibres varies as the winding angle of the cellulose (Fig. 3.8 which is for wet cellulose). Wetting the cellulose causes the stiffness to drop by a factor of 2 to 4 as the water penetrates the amorphous regions (hence 'swellulose' – Gordon 1976) effectively removing the contribution to the stiffness of a large number of H-bonds. Mark (1967) has shown that these H-bonds hold the chains together so effectively that failure of cellulose fibres is due not to chains slipping past each other but to main chains breaking. The strength of cellulose increases with the number of residues in the chain up to about 2500. Below this length failure is presumably by chains slipping. Since natural cellulose molecules are

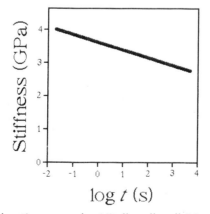

Figure 3.9 Stress-relaxation curve for *Nitella* cell wall (Houghton & Sellen 1968).

three to four times longer than this the H-bonds are, cooperatively, strong enough. This is probably true for chitin, though the highest chain length estimate is only about 700 residues (Ker 1977).

As might be expected, the stress-relaxation behaviour of cellulose is typical of a crystalline material (Fig. 3.9) in that it has a very broad spectrum (Houghton & Sellen 1968). A consequence of this is that the modulus and hysteresis of the material are virtually independent of strain-rate. This is more true for cellulose since its rate of relaxation is much slower than that of polyethylene or collagen, a consequence of extended chain conformation and the high degree of hydrogen bonding.

3.1.2 Gels

Cellulose is by no means the only polysaccharide of plant cell walls: there is a host of other substances, notably the pectins, which form lubricating gels or glues at various times in the plant's growth. But the mechanical properties of these materials are unexplored. However, there are some plant gels which have been very well investigated from the biochemical aspect and somewhat investigated mechanically. These are the carrageenans and alginates of commerce – gels extracted from dried seaweed – which are extensively used in the food, paint, clothing and other industries. These and the remaining polysaccharides to be discussed are formed of alternating residues which are

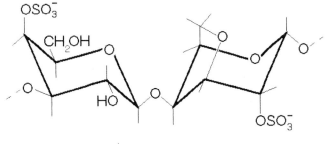

ι-carrageenan

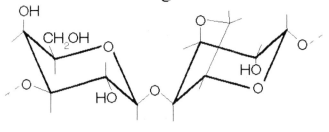

agarose

Figure 3.10 Alternating polysaccharides.

linked with alternating 1,3 and 1,4 links (Fig. 3.10). They are all capable of forming extended helices which intertwine or nestle within each other, forming structures which can be very stable in water, and thus have the characteristics of cross-links of greater or lesser strength. Gels so stabilized are permanent structures which have an elastic response to deformation. The formation of the double helices is problematic since it requires at least one of the molecules to have a free end so that it can wrap around its partner. A carrageenan might be 80% helical and the helical sections can be distributed in eight or ten zones along the molecule, each zone forming a double helix with a neighbouring molecule. Dai Rees always said that there was some equilibrium which allowed rearrangement of the helices so that they could form and unform quickly until all available helical segments were doubled with a neighbour, but admitted that this model required one of the chain ends to be free and capable of wrapping around the other chain! This idea seems dead now, but nothing has replaced it. One must assume that the observation of intercoiled rather than nestling helices is wrong. In agarose a number of such segments nestles together to give multiple cross-links; in alginates two chains nestle with the gaps filled with ions such as Ca^{2+} (Fig. 3.11; Mitchell & Blanshard 1974): the gelling properties of alginates are therefore partly dependent on the ionic conditions. Other polysaccharides seem to have a combination of double helical and nestling mechanisms (Fig. 3.12). All these gels, then, exist as lengths of cross-linked chain interspersed with lengths of more freely movable chain which bind and entrap vast quantities of water. For instance the average stable gel contains about 2% solid matter, but it is quite possible to have much more dilute gels which are much less stiff and yet coherent (Fig. 3.13; Mitchell & Blanshard 1974). At the other end of concentration, as a gel dries out it becomes much more stiff (Fig. 3.14; Denny 1984).

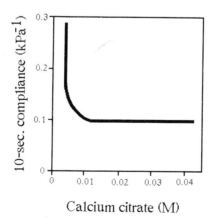

Calcium citrate (M)

Figure 3.11 The effect of Ca^{2+} (as citrate) on the 10-s compliance of 2% alginate gel in water at pH 6.0 (Mitchell & Blanshard 1974).

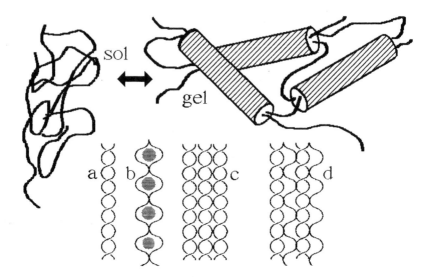

Figure 3.12 Stabilization of polysaccharides in gelling: a, by double helices (e.g. by l-carrageenan); b, bundles of double helices (e.g. agarose); c, 'egg-box' ribbon-ribbon interactions, Ca^{2+} sitting where the 'eggs' would be (e.g. alginate); d, mixed helix–ribbon association (Rees 1977).

The hydration of a gel is controlled largely by its polyelectrolyte nature: the sugar units carry negative charges which bind ions such as sodium or calcium. When there are only a few ions present the charges tend to repel each other and push the chain towards an extended conformation which entrains large amounts of water. This tendency is reinforced by the kinetic freedom of the chains and the associated ions, both of which will tend entropically to mix with the solvent as much as possible. Some water will be more or less bound, some

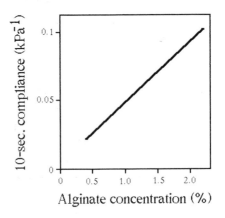

Figure 3.13 Effect of concentration on 10-s compliance of alginate gels formed at pH 6.0 with 0.035 M Ca^{2+} (as citrate) (Mitchell & Blanshard 1974).

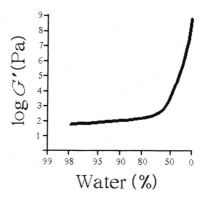

Figure 3.14 Shear stiffness of pedal mucus from *Ariolimax columbianus* increases as water content decreases (Denny 1984).

trapped. The ionic strength of pH of the solvent will obviously then affect the water content and stiffness of the gel. With alginates the stiffness decreases by an order of magnitude from pH3 to pH6, or by a factor of four as Ca^{2+} is reduced from 0.1 to 0.01 M.

A 2% elastic gel can be strained to 0.2–0.5 so it is not highly extensible (Fig. 3.15; Aizawa *et al.* 1973). Agarose and agar both show reasonable strength but are brittle. That is to say, a crack once started travels through the material quickly and easily. Table jelly is also brittle, and one only needs a spoon to break it. Kappa-carrageenan shows some ductility – there is some internal rearrangement which dissipates some of the applied force. Nuclear magnetic resonance (a method which gives information about the state of bonding of H^+ within a material) studies suggest that in brittle gels the water is more tightly bound whereas in a ductile gel the water is more free to move and this

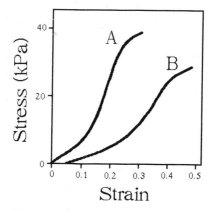

Figure 3.15 Stress–strain curves for agarose (A) and kappa-carrageenan (B) as 2% gels (Aizawa *et al.* 1973).

allows the chains to accommodate to changes in shape. The functions of such gels in the plant seem to be related to the stabilization of water around the plant and the ultrafiltration of ions. For instance agarose and carrageenan occur in different species of marine red algae as thick extracellular gels which maintain the plant's osmotic environment, give physical protection and permit and control the transport of metabolites between the plant and its environs.

3.1.3 The hyaluronic acid family

Whereas these plant gels have fairly stable cross-linking sites and are elastic, the related gels (Fig. 3.16) found in many animals must have less permanent cross-links since they are more viscoelastic and flow under stress. This means that the bonds must be made and broken and are presumably not as strong as those of the plant gels. At least in hyaluronic acid the molecules are double-helical, so perhaps the helical regions represent less of the molecule than in the plant gels. The other polysaccharides in Fig. 3.16 (known collectively as the polyuronides or glycosaminoglycans) are always associated with proteins and detail of their structure and functions is dealt with in Chapter 4. For the present,

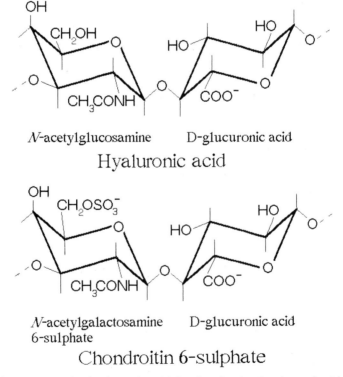

Figure 3.16 The hyaluronic acid family of animal polysaccharides.

the properties of the least-charged member of this group, hyaluronic acid (HA) will be considered.

Hyaluronic acid occurs in a comparatively pure form in synovial fluid, vitreous humour of the eye and in Wharton's jelly (contained in the umbilical cord). The molecular weight is probably in excess of 10^6 and may be as high as 10^7, depending on the source. This is probably a minimum estimate. Hyaluronic acid holds vast amounts of water in its structure and, like any gel, could be considered not as having a direct function such as load-bearing or lubrication, but as organizing the water so that the water itself is then better capable of load-bearing or lubrication. Although this may seem a philosophical point it is important, since it emphasizes yet again the necessity of considering biological materials in conjunction with their environment. To illustrate this, a single molecule of HA, molecular weight 10^6, would occupy in solution a sphere of diameter 1 m. The molecule itself would have a length of 2.4 m. Thus 1 g of HA would occupy 5 litres. Table 3.2 shows the degree of interpenetration of HA molecules at various concentrations.

Within the zones of overlap the interactions are not as permanent as they are in agar gels, as shown by the increased plasticity of HA. It seems likely that the helical cross-linking areas are continually changing to and from random

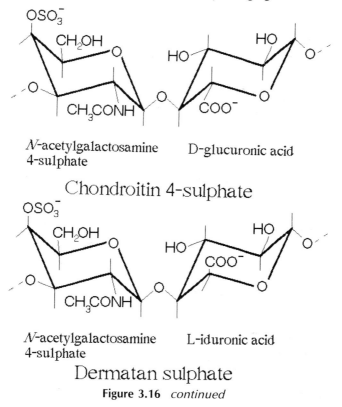

N-acetylgalactosamine   D-glucuronic acid
4-sulphate

Chondroitin 4-sulphate

N-acetylgalactosamine   L-iduronic acid
4-sulphate

Dermatan sulphate

Figure 3.16 *continued*

Table 3.2 The space-filling ability of hyaluronic acid

Hyaluronic acid (%)	Overlap of molecules (%)
0.02	none
0.1	80
0.5	96

coil conformation. This may well be happening in other animal polysaccharides, in balance between the forces of attraction and repulsion leading to a number of polymers with slightly differing properties for use in the production of different materials.

Hyaluronic acid shows a number of interesting rheological features seemingly the most important of which is that at low shear rates HA solutions are predominantly viscous, but are predominantly elastic at high shear rates. In itself this is not unusual, but it appears that there is virtually no transition region and that at high shear rates the G' and G'' curves diverge rapidly (Fig. 3.17; Gibbs *et al.* 1968). This suggests that the transition does not involve extensive relaxations. It may be that the system is heterogeneous so that small transitions can occur at different shear rates over a wide range thus effecting a smooth transition. The result of all this is most clearly seen and understood in the way viscosity changes with shear rate (Fig. 3.18). This shows that HA is a shear-thinning material and that as the shear rate is increased so the viscosity is reduced. This is also a property of certain non-drip paints. The relevance of this property to the functioning of synovial fluid in, say, the knee joint is not understood, although the reduced viscosity at high shear rates will allow the joint surfaces to move past each other with less loss of energy the faster the surfaces move. It seems likely that the most important factor in synovial fluid is its adherence

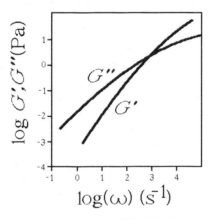

Figure 3.17 Loss (G'') and storage (G') moduli of hyaluronic acid. ω is the frequency of oscillation.

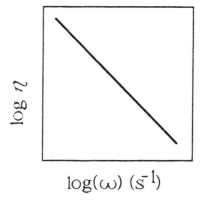

$\log \eta$

$\log(\omega)\ (s^{-1})$

Figure 3.18 Shear-thinning of hyaluronic acid: viscosity decreases with shear rate.

to the articulating surfaces to provide boundary lubrication, thus altering and controlling the surface properties of the articulation. How important HA is to this function is not known: presumably it is important and it is possible that one of the contributory causes of arthritis is the breakdown of the normal shear-thinning behaviour of HA in synovial fluid.

3.2 MUCUS

Mucus is a very strange material. The example shown – pedal mucus from gastropods – is rather more complex than a 'simple' polysaccharide gel, since it includes a certain amount of protein as well, though not in fibrous form. Such gels can have quite complex properties. For instance, hydrated pedal mucus of *Ariolimax columbianus* is a viscoelastic solid at small deformations (strains of less than 0.1), implying that the mucus is acting as a network. Thus at low rates of deformation (in sinusoidal oscillation, about 0.1 Hz) viscous effects predominate and there is some flow. At higher rates of deformation (10 Hz) the gel behaves like a rubbery solid. But if the gel is strained to 5 or 6, this network breaks down quite abruptly (yields) and the mucus becomes fluid. The yield strength depends upon, and is exponentially proportional to, strain rate, although the yield strain is independent of strain rate, again suggesting a viscoelastic network. Once the shear stops, the mucus heals quickly, the network reforms and the mucus becomes solid again. This solid–liquid cycle is used by the crawling slug as waves of contraction move across the sole of its foot. When part of the foot is moved forwards, the mucus becomes fluid and presents no resistance. In between the moving waves, parts of the foot are stationary with respect to the ground. Beneath these parts the mucus is solid. Denny (1984) calculated a reasonable crawling speed for the slug based on this model, suggesting that the rheological properties of the pedal mucus control the animal's

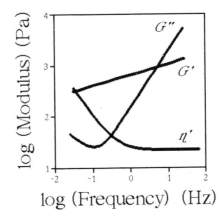

Figure 3.19 Viscoelastic characteristics of mucus from *Helix pomatia* (Simkiss & Wilbur 1977).

locomotion, or at the very least are tuned to the other factors involved (e.g. rate of contraction of the pedal muscles). The destruction of electrostatic interactions (e.g. by adding sodium dodecylsulphate (SDS), a detergent which gives the macromolecules a uniform negative charge) changes the rheological properties markedly. Fresh mucus from *Helix pomatia* shows similar behaviour to that of *A. columbianus* (Fig. 3.19; Simkiss & Wilbur 1977). At low rates of shear the mucus is more solid with a higher viscosity; the viscosity falls as the shear rate increases, finally becoming Newtonian (i.e. viscosity independent of shear rate) and the mucus becomes more fluid. However, the addition of sodium dodecylsulphate (SDS), of EDTA or of prolonged storage is to change the behaviour drastically (Fig. 3.20) and reduce both viscosity and moduli. Sodium dodecylsulphate, for instance, eliminates all electrostatic interactions

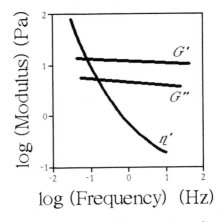

Figure 3.20 Same as Fig. 3.19, with molecular interactions disrupted by SDS (Simkiss & Wilbur 1977).

within and between molecules. This confirms the view that mucus can be considered as being made of a transient network that can be formed by the adherence of individual molecules at widely separated points (hence giving a low stiffness). If the mucus is allowed to dry out and the water content goes below 70%, it shows an abrupt increase in shear modulus (i.e. the water is thus acting as a plasticizer and the mucus is starting to go through the viscoelastic transition zone) getting up to about 50 MPa at 10% water content (Fig. 3.14). The dry material (possibly not totally dehydrated) has a shear modulus of about 1 GPa (the glassy modulus) but still has viscoelastic characteristics. In this condition it makes a very effective glue, although neither theory nor experiment are sufficiently developed to decide whether the effectiveness of adhesion limits the size of littoral organisms exposed to waves (Denny 1984).

3.3 THE COMPLIANT MATRIX

Shear-thinning may be a general feature of pliant biological materials, and the reason for thinking this (apart from the fact that it has occasionally been observed and commented upon) is based upon the mechanical properties of filled rubbers – rubbers to which have been added varying amounts of finely divided stiff material. Such mixtures are, in some respects, mechanically very similar to pliant composites such as are dealt with in Chapters 4 and 5, and to fluids such as mucus, synovial fluid and saliva. Many of the pliant biological composite materials can be considered as relatively stiff fibres or particles in a compliant matrix – the filled rubber equivalent has finely divided carbon black, up to 50% by volume, which is tightly bonded into the rubbery matrix. Other sorts of filler particle have been used, such as starch xanthide, glass beads, nylon, steel or cotton fibres, but in all instances the greatest filler effect is noted with fillers which bond well into the matrix. Other compliant matrices have been used and some gels, paints, inks and aqueous suspensions of particles (as in paper-making) have similar mechanical properties. It is now fairly well accepted that the polymer chains of the matrix in a strained filled rubber deform more in the region around a filler particle than in the region remote from the particle. This is called strain amplification; it puts extra stress on the interactions between filler and matrix and on interactions within the matrix immediately surrounding the filler particle. It is possible that, even at quite low overall strains, the strain amplification around a filler particle leads to breakdown and reorganization of these local interactions. These processes absorb energy so that the material as a whole exhibits hysteresis (Fig. 3.21), the area of the hysteresis loop representing the work lost in these internal rearrangements. Some of these rearrangements will be permanent, some temporary; some will recover in a short time, some will take longer, the time depending, as with all viscous processes, on the temperature. So, if the material is deformed a second time to the same extent as the first, two effects are noted: the second curve does not follow the first

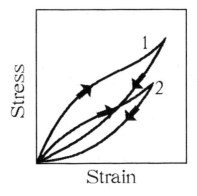

Figure 3.21 Hysteresis and stress-softening of a filled rubber over two cycles of straining.

one (the material is overall more compliant) and the area of the hysteresis loop is much reduced (Fig. 3.21, curve 2; Harwood *et al.* 1965). This effect (stress-softening or the Mullins effect) seems to occur in all reasonably compliant biomaterials so far investigated. It is usually regarded as a nuisance since repeatable stress–strain curves are impossible to obtain from a single specimen. The normal way in which this problem is overcome is to extend and retract the specimen many times (100 or so) until the hysteresis loops are repeatable (i.e. all the rearrangements with long relaxation times have been performed) and to call the material 'conditioned'. Although this is experimentally convenient, since it allows a single specimen to be used for several experiments and the results to be compared, it also means that the material is unlikely to be in the same mechanical state as it is in the living organism. There are also procedural problems such as the recovery of the processes with long relaxation times. You cannot wait very long between experiments or the material might need conditioning again. And does the specimen then suffer from mechanical fatigue? Perhaps arteries may be conditioned and give valid results, but it seems unlikely that many other biomaterials can be so treated and still be expected to give results which mean anything. An example is the extensible intersegmental membrane of the female locust for which stress-softening is functionally basic (Section 4.5). Not all soft composites show such extreme hysteresis and the degree of hysteresis probably depends to some extent on the shape of the stress–strain curve (Fig. 3.22). The fact that the added filler tends to increase hysteresis suggests that the filled compliant composite will have a high tan δ and be near a mechanical transition. In fact the tendency is for added filler to depress the peak of tan δ and for the plateau region to be extended. But at the same time the level of tan δ in the plateau region is increased (Fig. 3.23; Payne 1965). This implies that even though the addition of filler makes the elastomer stiffer (increases G') it increases the importance of viscous processes (G'') even more. The overall result is that the material is stiffened, strengthened and toughened by the addition

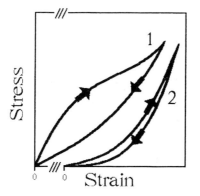

Figure 3.22 Comparison of hysteresis cycles of a rubber (1) and a typical collagenous biomaterial (2).

of filler. The relevance to biomaterials is not total, merely suggestive, since the fillers used in artificial elastomers are very stiff, whereas biomaterials such as skin and artery have components such as elastin whose compliance is similar to that of the composite but changes in a different manner with strain. Consequently one may speculate that the filler effect of elastin in such materials may be in some measure dependent on strain. But one important implication for biological materials is that stress-softening and other attributes of filled rubbers are reliant on the difference in modulus of the components. Several papers have been published in which specific components or specific morphologies have been cites as the cause of what is most likely just a filler

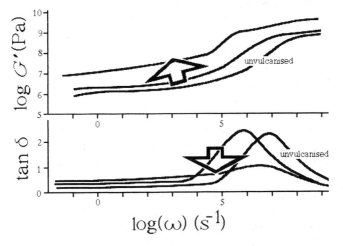

Figure 3.23 Master curves for elastic modulus (top) and loss angle (bottom) for styrene-butadiene rubber (SBR): note the effect on the pure vulcanized (i.e. cross-linked) SBR of adding 50% carbon black (arrow); compare with unvulcanized (Payne 1965).

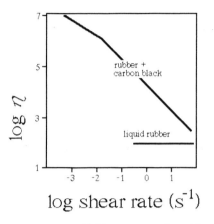

log shear rate (s^{-1})

Figure 3.24 The effect on viscosity of adding 50% carbon black to an uncross-linked liquid rubber (Berry & Morrell 1974).

effect due to the presence of a stiff component such as collagen (e.g. Serafini-Fracassini *et al.* 1977).

Another aspect of filled elastomers which seems to be rarely appreciated but which is of great importance to biological materials and must have an important effect on the dynamic functioning of the material, is the effect of fillers on the viscosity of the elastomer. In filled rubbers the storage modulus, G', has an extended plateau region over which it changes very little even though the strain rate may vary by orders of magnitude. This is an oft-quoted feature of soft biomaterials and has been noted in (amongst others) artery, skin, *Metridium* mesogloea, egg white, saliva and mucus. But if G' is relatively constant, the material is not in a viscoelastic transition zone, so G'' will also be more or less constant. Under these circumstances nu will be inversely proportional to strain rate (Eq. 1.29). This is most clearly shown in an experiment in which 50% carbon black was added to an otherwise Newtonian liquid rubber (Fig. 3.24, Berry & Morrell 1974). The resulting material has a high viscosity at low strain rates so that it shows no tendency to flow under its own weight. At higher strain rates the material has a much lower viscosity (note that in Fig. 3.24 the scales are logarithmic); just like hyaluronic acid it shows shear-thinning. Thus the material retains its shape at rest but can be readily transported and transformed under the action of relatively small forces. The importance of this fact seems not to have been realized in the study of soft biological materials, many of which presumably will show this effect since their stiffness varies so little with strain rate. Constancy of modulus with strain rate with the associated shear thinning also occurs in a number of mucilages and mucous substances such as saliva and cervical mucus. The resemblance to a filled elastomer may then be due to the breaking and remaking of these interactions under stress. This would mimic the rearrangement of interactions around the filler particle as the material is strained.

CHAPTER 4

Soggy Skeletons and
Shock Absorbers

Proteins and polysaccharides are the major classes of polymer used in biological materials. The number of materials which contain just protein or just polysaccharide is very small. By far the greatest number of biological materials contain both protein and polysaccharide, both more or less hydrated. The protein and the polysaccharide and the water can then be spoken of as phases. In most biological composites there is a fibrous phase of collagen and a matrix phase of other protein and polysaccharide. Defined in this way the matrix is the relatively amorphous material in which the fibres sit. Unfortunately the term 'matrix' is also used to denote any extracellular materials which form a supportive structure. This use of the term ('the extracellular matrix') will not be made again here: 'matrix' will refer to the material in which the fibres or particles of a composite material are embedded.

4.1 MIXING PROTEINS AND POLYSACCHARIDES

The vast number of soft, pliant tissues in the animal world are made of collagen fibres in a protein–polysaccharide matrix. And just as both collagen and poly-saccharides achieve various conformations according to their chemistry and their chemical environment, so it seems do mixtures of collagen and polysaccharides. For instance the addition of chondroitin 4-sulphate (C4-S) and chondroitin 6-sulphate (C6-S) (Fig. 3.16) to tropocollagen in physiological saline at pH 7.0 and 37°C will accelerate the formation of collagen and result in thinner fibrils. This and similar interactions seem to be largely due to ionic interactions between the positively charged lysine and arginine side chains of the protein and the negatively charged sulphate and carboxyl groups of the polysaccharide. The interactions can be broken down by high concentrations of salts and occur to differing extents according to the proteins and polysaccharides involved. The interactions are believed to control the size, orientation and rate of formation of collagen fibrils and so to play an important part in determining the mechanical properties of connective tissue.

Table 4.1 Acid mucopolysaccharide specimens

		Source	Mol.wt	Sulphates per disaccharide
Chondroitin-6-sulphate	C6-S	Umbilical chord	40 000	0.98
Chondroitin-4-sulphate	C4-S	Sturgeon notochord	12 000	0.97
Dermatan sulphate	DS	Pig mucosal tissues	27 000	1.29
Hyaluronic acid	HA	Umbilical chord	230 000	—
Heparan sulphate	HS	Cow lung	—	0.99
Keratin sulphate	KS-1	Cow cornea	16 000	1.17
Heparin	HEP	Pig mucosal tissues	11 000	2.33

From Blackwell & Gelman (1975)

Using a variety of polysaccharides (Table 4.1) and three homopolypeptides – poly(L-lysine), poly(L-arginine) and poly(L-ornithine) – as protein models, Blackwell and Gelman (1975) investigated the nature of such interactions. Using very dilute (0.0005 M with respect to the monomer residues) solutions, mixtures of the polypeptides and polysaccharides were made and the conformation of the mixtures investigated using circular dichroism (CD) spectroscopy. CD is a standard technique which is particularly useful for detecting the presence of alpha-helical material.

The spectrum obtained from the mixture is markedly different from that obtained by adding the spectra from the individual components indicating that interaction has occurred and that the conformation of one or both of the components has changed. Since the polysaccharide spectrum in the area studied (200–240 nm) is fairly constant and is relatively unaffected by changes in pH, salt concentration or temperature, it is reasonable to assume that it is the protein which has changed. If the CD spectrum for the polysaccharide alone is subtracted from the spectrum of the interacting components, the resulting difference spectrum will be for the protein alone. These spectra show that there is a significant additional proportion of α-helical conformation in the homopolypeptide. With C6-S as the polysaccharide the maximum amount of α-helical conformation is induced in the poly(L-lysine) at a ratio of 1:1, i.e. one lysine per disaccharide. Since de-sulphated C6-S induces no α-helical conformation in poly(L-lysine), it seems that the sulphate groups are interacting with the lysine side chains. If poly(L-arginine) is used instead of poly(L-lysine), the ratio at maximum interaction is 2:1, indicating that both the sulphate and the carboxyl groups of C6-S are interacting with the arginine side chains.

Table 4.2 shows the stoichiometry of several mixtures of polypeptides and mucopolysaccharides and also shows that the polypeptide is not always in α-helical conformation at the ratio of maximal interaction. Also shown in this table are the temperatures at which the interactions break down. These temperatures are quite well defined to within a degree or so; this sharpness of melting temperature is typical of non-covalent co-operative interactions. It is also a measure of the strength of the interactions: the stronger the interactions,

Table 4.2 Polypeptide/polysaccharide stoichiometry, conformation and melting temperature at maximum interaction: a, α-helix; R, random coil

Polypeptides		Polysaccharides						
		HA	C4-S	HS	C6-S	KS−1	DS	HEP
	ratio	1:1	2:1	1:1	2:1	1.2:1	1.4:1	3.3:1
poly(L-arg)	conf.	a	a	a	a	a	a	a
	$T_m(°C)$	35.0	54.5	65.0	76.0	90+	90+	90+
	ratio	1:1	1:1	2:1	1:1	1.2:1	1.4:1	2.3:1
poly(L-lys)	conf.	R	a	R	a	R	a	a
	$T_m(°C)$	—	25.0	—	47.5	—	76.5	90+
	ratio							2.3:1
poly(L-orn)	conf.							a
	$T_m(°C)$							56.0

From Blackwell & Gelman (1975)

the greater the degree of thermal agitation which can be tolerated before the interaction breaks down. Thus the mucopolysaccharides in Table 4.2 are listed in order of increasing strength of interaction. In summary, the stoichiometry and strength of the interactions between polypeptide and polysaccharide are dependent upon:

(a) the length of the amino acid side chain;
(b) the number of charges per disaccharide, especially the degree of sulphation;
(c) the position and orientation of the sulphate and carboxyl groups on the polysaccharide backbone.

Some of the information from this relatively simple model system can be used to gain insight into the more complex system created by replacing the homopolypeptide with collagen. At the melting temperature of the collagen (observed by CD) the collagen triple helix will be disrupted. Does the melting temperature change if the collagen is allowed to interact with the different polysaccharides? The answer which Blackwell and his colleagues found is yes. The change in temperature is not gradual but quantal, suggesting that the collagen either is or is not reacting totally with the polysaccharide (Fig. 4.1). Just as in the homopolypeptide model the interaction with different polysaccharides varies; Table 4.3 shows the number of disaccharide units

Table 4.3 Percentage disaccharide units needed to interact with amino acids to produce the complete transition of figure 4.1

HA	C4-S	HS	C6-S	KS—1	DS	HEP
11	14	—	5.5	10.0	5	—

From Blackwell & Gelman (1975)

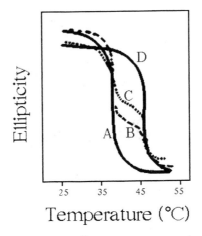

Figure 4.1 Melting curves for the collagen component of mixtures of collagen and chondroitin 6-sulphate; A, no polysaccharide; B, 1% polysaccharide residues; C, 3% polysaccharide residues; D, 5.5% polysaccharide residues (percentages are as polysaccharide residues relative to amino acid residues) (Blackwell & Gelman 1975).

per 100 amino acids required to produce a complete change in the melting curve. In each instance the collagen has all been stabilized and in each case the melting temperature is 46°C to within a degree or so. So although the relative amounts of the components needed to form the cooperative interaction vary with the components, the nature of the stabilization is probably very constant and similar.

As mentioned above, the addition of mucopolysaccharides to tropocollagen in solution affects the formation of collagen fibrils: Blackwell's experiments lend further weight to the idea that such interactions are a controlling factor in the aggregation of collagen fibrils in native connective tissues. However, the exact interactions which occur in collagenous tissues and the effect which they have on the stability and composition of tissues are rather more equivocal. This is possibly because it is very difficult to separate the components without contamination of one with the other and partly because in the various experiments which have been performed using components derived directly from collagenous tissues the results have been very varied. It appears that temperature, ionic strength, pH, source of the collagen (i.e. its exact chemistry) and the molecular weight distribution of the polysaccharide are all important variables. It is to be expected that various tissues have these variables controlled by the cells producing them, thus ensuring a material which is presumably nicely adapted to its function. The complexity of the interactions is further increased by the presence of non-collagenous protein linking polysaccharide chains into larger units. This ultimately is currently represented by the 'lampbrush' model (Fig. 4.2). But, however one depicts the arrangement of the components, the mechanically important factors will include the following.

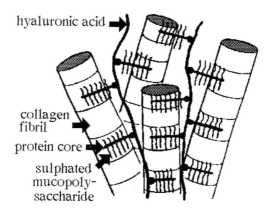

Figure 4.2 Collagen-proteoglycan-hyaluronic acid model of soft tissue. The proteoglycan is in 'lampbrush' or 'bottlebrush' conformation (Hukins 1975).

1. The stability – and hence, presumably, stiffness and strength – of the collagen is increased by the interactions.
2. The fibrils are stabilized by extensive interactions within a protein–polysaccharide matrix. The material is therefore a composite and such factors as the shear stiffness of the matrix (which will affect the transmission of loads within and through the material), water content, volume fraction of the components, orientation of the fibrils and length of the fibrils will be important in controlling the mechanical properties of the matrix.
3. If the interactions produce maximum stability of the components at a defined ratio between the amounts of the components, then it seems possible that the biochemistry of the components will greatly affect the volume fractions of the components and the completeness of their interactions, suggesting that the biochemistry of the components can be more or less directly linked to the mechanical properties of the entire interacting material (see above).

4.2 STRESS–STRAIN BEHAVIOUR

Most soft tissues can be modelled as soft composites of an elastomer filled with fibrous or particulate material (Jeronimidis & Vincent 1984). Mesogloea and most soft insect cuticles (such as maggot cuticle, but not locust intersegmental membrane which has the chitin in highly orientated order) are both examples of this and show a similar sort of stress–strain curve (Fig. 4.3). Skin, artery, gut wall, bladder and nearly all other collagenous tissues also show this sort of stress–strain curve. It is usually interpreted as follows: the first, low modulus, region is the elastin stretching, the high modulus region is due to the collagen; the collagen contributes relatively little to the low modulus region because it

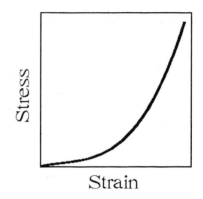

Figure 4.3 Stress–strain curve of a collagenous biomaterial.

is crimped or folded or just arranged randomly. Apart from the fact that this explanation cannot work for soft insect cuticle which contains no elastin, there are other ways of obtaining this sort of stress–strain curve which may give some hints of a rather more general model. A familiar example is knitted or woven fabric (the latter pulled at 45°C to the warp and weft). When either material is stretched the increasing resistance to extension is caused by the progressive orientation of the fibres until all the fibres are orientated in the direction of extension and the modulus has increased to that of the fibres. This gradual increase in the modulus (the incremental modulus) will give the same sort of stress–strain curve as Fig. 4.3. For comparison with this model, the orientation of collagen in the wall of the aorta of the pig (which is composed of layers of

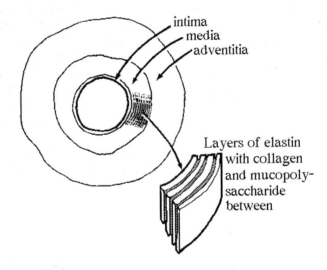

Figure 4.4 Basic morphology of an artery wall.

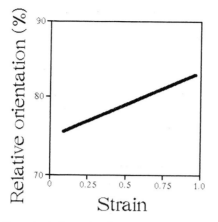

Figure 4.5 Change in the orientation of collagen (expressed as a function of angular dispersion) with strain.

elastin and randomly orientated collagen, Fig. 4.4) has been followed by X-ray diffraction while the artery was being extended. The mean orientation of the collagen can be deduced from these results to show how it changes with strain (Fig. 4.5). Considering that this is a statistical distribution of collagen orientation it will be apparent that some of the collagen fibres will be stressed even at low strains. So then the progressive increase in orientation of the collagen is effectively an increased recruitment to a population of load-bearing fibres aligned sufficiently to support the applied stress and hence the 'modulus' of the material (or is it a structure?) (Aspden 1986, 1988). Two salient facts emerge on consideration of this model: first, that even at low strains it is possible that collagen fibres can be fully extended and can be broken. This means that the material is breaking down in a minor sort of way even at low strains; it also means that mechanisms for toughening the material – dissipating the strain energy by limited breakage of the bonds – are available at small strains. Compare this with the effect of a filler on toughening rubber (section 3.2). The second fact is that at any one time, even at high strains, by no means all the collagen fibres are taking the load: in fact probably only 40% or so of the fibres are load-bearing at maximum stress, so that the material can go on being load-bearing even after it has started to break down. This all means that any model of soft tissues must take into account not only the amount of collagen present but also the effect of strain on re-orientating the collagen (Aspden 1986; see also Section 4.6).

4.3 POISSON'S RATIO

Further spanners can be thrown into this wheel! Some indication of the relative importance of the elastin and collagen components of skin can be gained from studying Poisson's ratio. The principal Poisson's ratio is the ratio between the elongation (the applied strain) and the lateral contraction (in either width or thickness – the induced strain) of the specimen (Fig. 4.6):

$$\nu_{xy} = \frac{-e_y}{e_x} \qquad \text{[Eq.4.1]}$$

For an anisotropic material (which includes nearly all biological materials, since so many are composites with a preferred orientation) there are six Poisson's ratios (Jeronimidis & Vincent 1984). But for present, consider only the second Poisson's ratio in the plane of a sheet of anisotropic material. There must be two, different, Poisson's ratios because:

$$\frac{\nu_{xy}}{E_x} = \frac{\nu_{yx}}{E_y} \qquad \text{[Eq.4.2]}$$

If the material is incompressible then the two ratios add up to unity. This obviously will apply to any pair of ratios within the material. Since $E = \sigma/e$, strain can be substituted in [Eq.4.1] to give:

$$e_x = -\nu_{xy}\frac{\sigma_y}{E_y} \qquad \text{[Eq.4.3]}$$

If the material is being stretched in two directions at once, then:

$$e_x = \frac{\sigma_x}{E_x} - \nu_{xy}\frac{\sigma_y}{E_y}; \; e_y = \frac{\sigma_y}{E_y} - \nu_{yx}\frac{\sigma_x}{E_x} \qquad \text{[Eqs 4.4a,b]}$$

In a tube under pressure, where the hoop stress will be twice the longitudinal stress, these two equations can be combined, but only if the material is isotropic, to allow calculation of Poisson's ratio:

$$\frac{e_x}{e_y} = \frac{1 - 2\nu_{yx}}{2 - \nu_{yx}} \qquad \text{[Eq.4.5]}$$

Otherwise Poisson's ratio has to be measured independently in the two directions.

Poisson's ratio is often neglected: it usually appears as a small, constant number in many equations concerning biological materials and, as often as not, its value is assumed to be 0.5, which is the value for rubbers at small extensions (Fig. 4.7) and which assumes that the material retains a constant volume. The Poisson's ratio of rubber drops only if engineering strain is used for the calculation. If the material is isotropic and constant volume, use of true strain in the calculation of Poisson's ratio will give a steady value of 0.5 independent

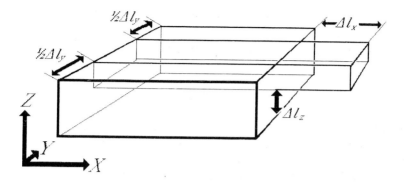

Figure 4.6 Conditions for the definition of Poisson's ratio.

of the degree of stretching. Problems occur partly because soft tissues are extended by relatively large amounts (strains of 0.5 and more), partly because they contain fibres and partly through confusion as to which form of strain is the more appropriate to use for the calculation. For instance a Poisson's ration of 1.0 can be obtained from a network or trellis work such as an orthogonally woven cloth (e.g. a handkerchief) but only in directions at 45° to the warp and weft. The skin on your belly has similar properties (JE Gordon, personal communication). A high Poisson's ratio in all directions is characteristic, if not diagnostic, of an open feltwork rather like a haystack. It is also quite feasible to have an open feltwork which is embedded in a soggy jelly which will flow in and out of the mesh, whose Poisson's ratio is 1.0 or greater, yet whose volume is constant or nearly so. Thus the assumption that because biological materials contain water and are therefore incompressible it therefore follows that their Poisson's ratio is 0.5 is not tenable (Jeronimidis & Vincent 1984). Not only is it possible to have voids in a material filled with 'alien' fluids but it is also possible

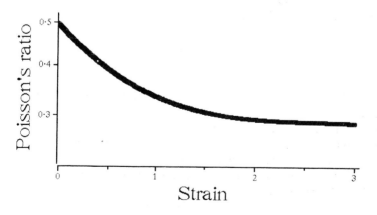

Figure 4.7 Variation of Poisson's ratio with strain in a rubber.

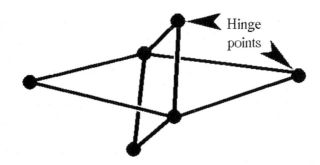

Figure 4.8 A strut framework which will give strange Poisson's ratios.

to have a high Poisson's ratio and constant volume or a vanishingly low Poisson's ratio with a feltwork of fibres embedded in a viscous matrix. And once you start thinking of different ways of putting a material together it soon appears possible (using strut frameworks resembling the fibres in a feltwork, Fig. 4.8) to generate Poisson's ratios of all magnitudes, even negative ones in which the material *expands* in one of the directions orthogonal to the extension. This entertaining phenomenon can be observed in certain strange open-cell foams and networks which have been squashed plastically and then deform elastically (Lakes 1987, Gibson & Ashby 1988). But until a proper study is made of Poisson's ratios of soft tissues under the varying states of initial strain, ν will remain not merely neglected but inscrutable (Jeronimidis & Vincent 1984). Worse still, E and G cannot be related with any certainty using Eq. [1.5]: indeed the true stress and strain which should be used for extensions greater than 0.1 cannot be compressed even into a simple formula for calculating E, so that much of biomechanics theory (many equations concerning the mechanics of skin and artery) is wrong since the assumption '$\nu = 0.5$' has been made (but see below).

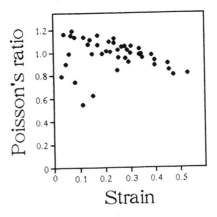

Figure 4.9 Poisson's ratio of the skin from a complete cow's teat (Lees 1989).

The principal Poisson's ratio of cow teat skin, stretched as a cylinder by internal pressure (Fig. 4.9; Lees 1989) is totally different from that of rubber. The teat was lined with a thin rubber sheath (left slack so that it didn't affect the expansion of the teat), mounted on a brass support and inflated with air. The fiducial marks were indicated by very small spots of ink. The specimens were inflated by small amounts, photographed and the measurements taken from photographic enlargements. It looks as if cow teat skin is in the 'open feltwork' category, which is not surprising when you examine sections of the teat: there are many concentric layers which are permeated with blood vessels and spaces of various sorts. Experiments with isolated strips of teat skin of aspect ratio varying between 1.5 and 10 gave very different results with negative Poisson's ratios when the aspect ratio of the test piece was below 2.5. With a relatively short specimen, it was getting wider as it was being stretched. However, one of the variables was lateral prestrain at the clamps – the skin is very highly folded, and the act of gluing it on to an aluminium tab (which will be gripped by the clamps of the test machine) flattens out these folds.

Aorta, also inflated with gas but prestrained as in life, gave a very different result (Fig. 4.10; Dobrin & Doyle 1970). But in this instance Poisson's ratio was calculated from the ratio of the slight, sequential, increases in longitudinal and circumferential stress as the artery was inflated. Although it is not very clearly stated, it is probable that the stresses were calculated from the instantaneous rather than the original thickness of the artery wall. Gosline & Shadwick (1982) use the same model to derive a principal Poisson's ratio for the untethered arteries of certain cephalopod molluscs as slightly under 0.5. They definitely use the instantaneous cross-sectional area of the specimen in their calculation of modulus, so their results are calculated using the true stress and are therefore more easy to apprehend than the cow teat data. A principal

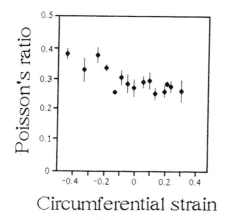

Figure 4.10 Poisson's ratio of prestretched carotid artery from a dog (Dobrin & Doyle 1970).

Poisson's ratio of 0.5 is typical, though not diagnostic, of a material elongating at constant volume.

With all three of these sets of results only one out of a possible six Poisson's ratios is presented. So there is no information about changes through the thickness of the specimen. And there is insufficient information from these results to be able to deduce anything about changes in thickness. Experimentally, it is very difficult to measure the thickness of pliant materials like skin. It is also possible that the membrane is actually reducing in volume during the experiment and that holes in the thickness of the material such as capillaries are being closed on themselves. This may be why the skin over your knuckles goes white when you stretch it by clenching your fist. A rope will behave in rather similar fashion: if the rope is wetted and then stretched much of the water will be forced out from between the fibres as the volume of the rope decreases. This is because the rope is a fibrous system and orientation of the fibres causes an increase in Poisson's ratio. So although Poisson's ratio is an important parameter, it has to be handled with extreme care.

If the fibres in the material are orientated orthogonally to the direction of stretching (Fig. 4.11), any amount of stretch will not reorientate them (in the perfect example) but will cause them to change their position relative to one another. In this instance Poisson's ratio will be very small, measured in the plane of the material (Fig. 4.12) and the material will deform much through its thickness: the locust extensible membrane does this and has a Poisson's ratio of 0.1–0.02 in the plane of the material, so the diameter of the abdomen of the digging locust does not change very much but the membrane gets very thin as it is being stretched. One extra point to notice concerns the analysis of the way *Metridium* expands itself. The explanation given below is that the different stiffnesses interact with the different stresses set up in the cylinder of mesogloea and produce a change in shape of the animal such that it elongates rather than swells laterally. This is the other side of an argument which could be approached from Poisson's ratio of mesogloea. This ratio has not been measured for mesogloea, but the non-random orientation of fibres is going to have similar effects to those in locust membrane, where the strain in the direction of the fibres is appreciably less than the strain orthogonal to the fibres when the material is stressed. Thus the outer layer of mesogloea (Fig. 4.13) will tend to deform more through its thickness (get thinner as it is stretched). Its Poisson's ratio in the length/circumference plane will be about 1.0 whether it is strained longitudinally or circumferentially. But the inner layer will have a very low Poisson's ratio when stretched longitudinally. How these two ratios will affect each other is unknown, but it is likely that mesogloea is not the only soft tissue which has layers in which the fibres are orientated differently.

This discussion of Poisson's ratio in fibrous biological materials has done no more than introduce some of the complexities. Once again, the theory of deformation at high strains just does not exist. Probably a large part of the

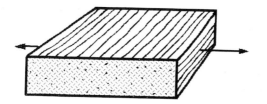

Figure 4.11 Orientation of the fibrous (chitin) phase in locust intersegmental membrane cuticle relative to the direction of extension.

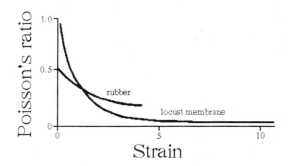

Figure 4.12 Poisson's ratio of locust intersegmental membrane cuticle (Fig. 4.11) compared with that of rubber, an isotropic material.

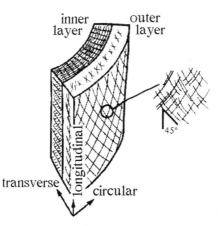

Figure 4.13 A section of the mesogloea of *Metridium senile* showing the orientation of collagen fibres in the two layers.

seeming inscrutability of such parameters as Poisson's ratio is due to this lack. Until we have found a workable way of dealing, mathematically, with materials at high strains we shall continually be discovering incongruities which are more apparent than real. One approach to this which has not yet been tried, as far as I know, is to adapt computer models developed for designing tensile structures such as tents. These use a combination of finite element analysis and iterative calculation involving reorientation functions to model woven materials. The higher strains observed in soft tissues would need a more sophisticated approach, but could in principle be analysed in a similar fashion.

4.4 THE SKELETON OF THE SEA ANEMONE

An example of a simple connective tissue composed of discontinuous collagen fibres in a mainly polysaccharide matrix is the mesogloea of the sea anemone (Gosline 1971). In one sample of mesogloea taken from eight individuals of *Metridium senile* collected off the coast of California, Gosline found that the composition was as follows:

salt – 5%; collagen – 6.7%; matrix – 2%; water – 86%

The matrix material was found to be mainly neutral hexose polysaccharides plus some protein. Making allowance for the water bound to the collagen it turns out that the matrix is a gel of 2.4% solids content. This is well within the normal range of concentration for gels (e.g. an agar gel of 2% is very rigid; gelatin gels of 1% are stable, see also Fig. 3.13). The collagen has similar swelling and birefringence behaviour to rat tail tendon collagen and gives a similar X-ray diffraction pattern. Mesogloeal collagen, when compared with that of rat tail tendon (Table 4.4; Brown 1975), contains rather less proline and hydroxyproline but is otherwise similar. The assumption is therefore made that the mechanical properties of the two collagens are also rather similar, with a modulus of around 1 GPa and an elastic strain limit of about 0.04. The mesogloea of *Metridium* strains to 3 or more, so it is clear that the collagen in the matrix is discontinuous

Table 4.4 Collagen amino acid content: rat-tail tendon (RTT) compared with collagen from the mesogloea of *Metridium senile*

	RTT	*M. senile*		RTT	*M. senile*
Ala	9.9	11.3	Met	0.6	0.9
Gly	35.1	30.8	Arg	4.7	5.7
Val	2.3	3.4	Hist	0.3	0.5
Leu	2.2	3.7	Lys	3.6	2.7
Ileu	1.3	2.3	Asp	5.7	8.1
Pro	12.3	6.3	Glu	7.4	9.5
Phe	1.4	1.2	Hypro	9.0	4.9
Tyr	0.5	0.8	Hyly	—	2.5
Ser	2.8	5.4	Cyst	—	3.2
Thr	1.9	3.9			

From Brown (1975)

and that the main mechanical properties will be due to the matrix, probably modified by the collagen. This modification will be influenced by the length and orientation of the collagen and by the strain of the tissue: higher strains tend to orientate the collagen so that it bears more of the total load and the mesogloea stiffens in that direction. However, relaxation rate is as important as strain with a viscoelastic composite such as mesogloea. The relaxation modulus at short times is of the order of 0.1 to 1 MPa. But at long times the equilibrium modulus is more like 1 kPa. This makes it possible for the animal to stand up to short term forces such as surge yet inflate itself with the low pressures (1 Pa or less) generated by cilia of the siphonoglyph. When the load is removed from the mesogloea it recovers, in time, elastically. This suggests that the system is rubbery, which in turn invokes the presence of long chain molecules of high molecular weight. The conclusion that the essential mechanism is rubbery and that the material is working in the rubbery plateau of the viscoelastic spectrum is supported by the existence of a transition in modulus at about $-15°C$ at which temperature the storage modulus starts to increase, reaching about 100 MPa at $-22°C$. This transition is probably in part due to the action of ice crystals stiffening the matrix.

If the normal state of the matrix is rubbery, then the matrix molecules must have great freedom of movement to give such a low stiffness. This freedom seems to be given partly by making the matrix very dilute and so separating the matrix molecules. Greater freedom of the mesogloea matrix molecules is given by salts masking the charged sites on the sugar residues, thus preventing electrostatic interactions which would hinder rotation about the $-O-$ bonds of the polysaccharide backbone: if the mesogloea is washed in distilled water, so removing the salts (and, incidentally, a significant amount of mesogloea, according to Purslow 1980), the stiffness increases 30-fold; acid polysaccharides would be too highly charged and more likely to interact. They would also be less likely to form random coils and so less likely to give rise to a rubbery material.

The collagen in the mesogloea does not form a random feltwork; in *Metridium* it is distinctly organized (Fig. 4.13). The inner layer has a higher volume fraction of collagen than the outer layer so the collagen may be expected to have a greater effect on the mechanical properties of the inner layer. When the mesogloea is extended longitudinally the collagen fibres in the inner layer will have their orientation hardly affected by the strain and so will have very little effect on the modulus. The outer layer will have the lattice work more nearly orientated in the direction of straining and so will increase the modulus. At a strain of 0.4 or so this orientation will be more or less complete so there will be little further increase in modulus at higher strains. Since the matrix will now be shearing past the collagen the shear properties of the collagen matrix interface will become more important. Also it is unlikely that adjacent collagen fibres will be sheared in the same direction. This will cause the shear strain in the matrix to be greater than the overall shear strain of the material (Fig. 4.14). This shear

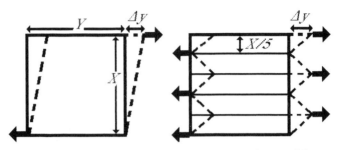

Figure 4.14 Effect of an additional reinforcing phase on the amplification of shear strain. top, $y_1 = dy/x$; bottom, $y_2 = dy/(x/5) = 5y_1$ (Wainwright *et al.* 1975).

amplification due to filler particles increases local shear strain and, because the shear modulus is not affected, increases the stress required to shear the entire piece of material. Thus the overall effect is to increase the stiffness of the material. This phenomenon is associated with fillers in general, whether they are fibrous or granular, and is at least partly the cause of the increase in stiffness when fillers are added to elastomers. When the mesogloea is extended circumferentially the orientation effect of the collagen in the outer layer is exactly the same as before. But the circumferentially orientated collagen of the inner layer will cause an increase in stiffness in this direction. In fact it is found that the circumferential stiffness is about three times the longitudinal stiffness. The significance of this is that the circumferential or hoop stress of a cylinder under pressure is twice the longitudinal stress (Fig. 4.15). So if the stiffness varies in the same ratio, the strains induced by any increase in internal pressure will be the same in the two directions. If, as in the mesogloea of *Metridium*, the hoop modulus is more than twice the longitudinal modulus then as the animal blows itself up it will extend away from the substrate at a greater rate than it increases in girth.

The creep behaviour of mesogloea has been investigated, albeit with an inappropriate analysis (Section 1.4.2). But even assuming the analysis were

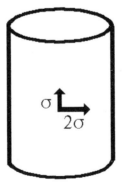

Figure 4.15 In a cylinder which is under internal pressure, the hoop stress is twice the longitudinal stress.

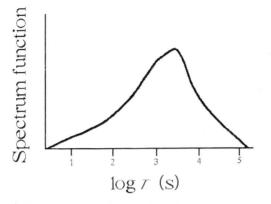

Figure 4.16 Retardation spectrum of mesogloea from *Metridium* (calculated from Alexander 1962 using the Alfrey approximation).

appropriate for such a high strain material the results are not as might be expected. The retardation spectrum (Fig. 4.16) shows that the distribution of retardation times is very discrete compared with a vulcanized *Hevea* rubber filled with 50 parts by weight of carbon black. The narrowness of the distribution also makes it rather risky to derive the spectrum with the Alfrey approximation, but be that as it may it seems highly unusual that mesogloea, modelled as a filled rubber composite, should have such a narrow retardation spectrum. Either the interpretation of the structure of the material is wrong or there is a single, very well defined, process controlling the mechanics. This problem clearly needs to be investigated further. One explanation mooted is that there are two polymeric systems in parallel, one cross-linked providing the modulus and one not cross-linked providing the viscosity, both with short retardation times. The steady-state elastic compliance of the 'viscous' system would appear as the instantaneous strain on loading. At times greater than the retardation times of its components such a double system would show, it is suggested, a very narrow spectrum (Alexander 1962). This does not explain why such a narrow spectrum is not seen in lightly cross-linked filled rubbers.

The relationship between the volume fractions of the components of the mesogloea of sea anemones and the mechanical properties is obscure. Within the species *Metridium senile* there is great variation: the salt-free solids can vary between 9 and 22% of the wet weight. Different areas around the coast of Britain yield animals whose mesogloea has varying mechanical properties but apparently fairly constant composition. Mesogloea of Scottish *M. senile* collected in the Firth of Clyde has more variable mechanical properties than that from English individuals dredged off Plymouth (Purslow 1980). Perhaps these variations are a reflection of the type of reproduction – sexual or asexual – which the animals mainly use in the different areas: members of a clone would be less variable in their characteristics. Alternatively it is possible that there are many sub-species of *M. senile*: how big is a single interbreeding population of this animal?

Problems like this are no less relevant to the biomechanical approach than to the more common physiological or ecological approach.

Some of the variation between species of sea anemones has been studied (Koehl 1977). *Metridium senile* lives in relatively sheltered waters; *Anthopleura xanthogrammica* lives in more exposed places. Thus whereas the maximum water velocity experienced by an *M. senile* might be less than $0.2 \, m \, s^{-1}$, the water velocity experienced by *A. xanthogrammica* may be two or three times that amount. Despite these differences, both experience similar drag forces of about one newton since *A. xanthogrammica* is much more squat and streamlined. Mechanical tests on the mesogloea of both species show that *A. xanthogrammica* has a far quicker elastic response after deformation: a force of about 30 kPa will extend mesogloea from both species to a strain of 0.6–0.65 so the instantaneous modulus is similar. But *M. senile* mesogloea creeps at a greater rate and when the load is removed it does not recover so rapidly; it still has a residual strain of 0.1 by the time that *A. xanthogrammica* has recovered completely. If the applied force (in this experiment applied for only a few seconds) be considered as analagous to a wave then *A. xanthogrammica* will not be much affected by wave action whereas *M. senile* will be further extended by each successive wave. If a force of 3 kPa is allowed to act on mesogloea of the two species for much longer times – of the order of a day – then *M. senile* will extend to a strain of 1.5, *A. xanthogrammica* to a strain of only 0.4 or so. In dynamic tests at frequencies between $10^{1.5}$ and 10 Hz the storage modulus of the two mesogloeas is the same, but *A. xanthogrammica* has a much lower tan δ, especially at the higher frequencies. This reflects the quicker recovery response of this mesogloea since it implies that more of the imposed deformation energy will be stored elastically.

To summarize, it seems that the major difference between the two mesogloeas is that *M. senile* has a greater viscous component. This could be achieved quite simply by controlling the permanence of the interactions between the components of the matrix and the fibres. If the interactions in *M. senile* mesogloea were relatively less stable, then the differences could be accounted for. Since the instantaneous modulus of both mesogloeas is the same then the short time interactions and the units contributing to them must be similar in both. The extreme difficulty of measuring composition, at least of *M. senile*, leads one to question severely the alternative hypothesis that there is a greater concentration of solids in *A. xanthogrammica*. The published comparisons of composition are based on very few analyses (eight for *M. senile*, three for *A. xanthogrammica*), and the *M. senile* analyses were not of those animals used in the comparative mechanical tests. But there are differences in the morphology of the mesogloeas. Notably *A. xanthogrammica* mesogloea has less well orientated collagen which is packed into much larger bundles and may therefore effect the more permanent cross-links demanded by the mechanical data. Another way of effecting more permanent cross-links would be to alter the chemistry of the polysaccharides. *M. senile* has mainly neutral polysaccharides so the possible interactions will be minimal. If the polysaccharides of *A. xanthogrammica* mesogloea were more highly charged then it would be reasonable to expect greater stability of the interactions between them. This work remains to be done.

Sea anemone mesogloea has been considered at some length for a number of reasons. The tissue is relatively simple, so it should be possible (though clearly not yet) to account for its mechanical properties in terms of its chemistry and morphology. The fact that this has not yet been satisfactorily achieved does not encourage one to have much faith in many of the models proposed for skin and artery, which are much more complex. It is probably also true that its function in the animal is relatively simple: it is mostly subject to only two sources of stress – internal pressure which acts slowly and external water motion which may be continuous (currents) or transient (waves). It is thus not unreasonable to suppose that it is possible to account for the mechanical properties of the material in terms of the demands of the animal's life style.

4.5 STRETCHING THE PREGNANT LOCUST

Another pliant material whose mechanical properties have successfully been related to the life of the animal possessing it is the highly extensible intersegmental membrane (i.s.m.) of the female locust (Acridiidae). The story is that the locust is a migratory desert-living animal and so has to take account of the problems of power/weight ratio (for maximum flight range) and of water supply in general. In practice this means that in order to reduce its flight payload the pregnant female carries her eggs in a relatively dehydrated state: they will absorb their own weight in water before they have developed into larvae. But in order to assure for her eggs a reliable supply of water she has to lay them deep in the ground (about 8 cm down) usually in the shade of a rock. To reach these depths she has a highly extensible i.s.m. which extends elastically to strains of 15.

The great problem for an animal with an extensible material which is elastic is for it not to be caught out by the recoil of the material when the force extending it is removed. For a long time this was thought to be the locust's biggest problem and some rather messy experiments purported to show that the locust was blowing its abdomen out like a balloon under internal pressure. But it was subsequently found, simply by making a small hole in the digging locust, that there was no internal pressure, so that the ovipositor assembly at the end of the abdomen must be (and indeed was) digging into the ground pulling the abdomen down after it (Vincent & Wood 1972). But what about the recoil problem? It turns out that although the locust does let go of the sides of the hole as it is digging (Vincent 1975b) it is safe from catastrophic recoil of its abdomen if the weight of the abdomen (about 1 g) can somehow hold the i.s.m. extended (it takes about 15 g to extend just the i.s.m., discounting the force needed to stretch the internal organs of the locust). And the way to do this is

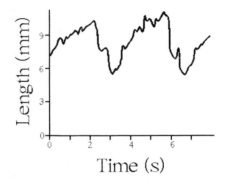

Figure 4.17 Change in length of locust intersegmental membrane while it is digging its oviposition hole (Vincent 1975)

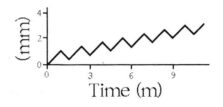

Figure 4.18 Straining program used on a tensile test machine to imitate the strains shown in Figure 4.17.

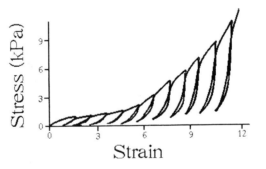

Figure 4.19 Stress–strain curve derived from the straining program of Fig. 4.18.

by stress-softening (Section 3.2). Figure 4.17 shows the strains that an i.s.m. undergoes during part of the digging cycle; Fig. 4.18 shows how these strains were modelled on a test machine and Fig. 4.19 shows the response of the i.s.m. to this straining programme (Vincent 1975a). The last cycle shows that with only a very small reduction in length the force required to hold the i.s.m. extended can drop by an order of magnitude – from 15 to 1.5 g. So there is no need for internal pressure to retain the extension. This experiment also highlights a general point about the significance of the mechanical properties of biological materials: a host of other experiments on the mechanics of the i.s.m. gave some interesting information about material, but it was not until the test mimicked the way the animal uses the material that the significance of the particular mechanical properties of the i.s.m. became apparent. Since biological materials are often designed with particular working conditions to accommodate, it is important to know what those working conditions are in order to interpret the mechanics with any degree of insight.

The i.s.m. is made of about 12% protein, 12% chitin and the rest is water. Despite the higher solids content the i.s.m. is much more pliant than mesogloea (about 5 kPa, about the same as sputum) and has a complex morphology which is probably designed to contain such a soft material and stop it from oozing away. The chitin plays no part in the elastic recoil properties since it is arranged at right angles to the direction of extension. The protein must be the elastomer. But it seems very likely that the chitin is important as a filler, much as the collagen in mesogloea, to encourage stress-softening and also to affect the Poisson's ratio. For the locust, stress-softening is a very important property, since it gives the insect a material which is stiff enough not to extend when the locust does not want it to, e.g. just with the weight of the abdomen hanging down, but is soft enough under certain circumstances to be held extended by that same small force. This i.s.m. recovers these characteristics between egg-layings, which occur only once every ten days or so. The abdomen is retracted after oviposition by super-contracting muscles.

4.6   FRACTURE

Another aspect of pliant biological tissues which does not receive much attention concerns how they break or tear. It is obvious that a bone has to be able to resist breaking and this is emphasized by the people who have to wear plaster casts and walk on crutches while their broken leg is healing. But soft tissues are just as liable to failure and will fail just as catastrophically. For instance the pilot of an airliner which crashed at Staines a few miles out of London airport in 1973 died as a result of his aorta splitting longitudinally. A sausage with a plastic skin will also split longitudinally when the meat swells up during cooking, and this is the preferred direction for a cylinder to split since, in a pressurized cylinder made of isotropic material, the hoop stress is twice the longitudinal stress. However, in both the aorta (of the pig) and in the body wall of the sea anemone, it is more difficult to propagate a crack longitudinally than circumferentially. So the airline pilot's aorta must have had some pathological difference in its structure to make it act like an isotropic material. Just what the difference might have been and what it was caused by we do not know. The factors affecting toughness in soft tissues are not very obvious. Table 4.5 shows the toughness of a number of soft tissues together with their stiffness. It is obvious that stiffness and toughness are not related. Partly this is because toughness measures the work put into a material to propagate a crack and since

Table 4.5 Toughness values for some soft biological materials

	Mean tearing energy (kJ m^{-2})	Stiffness (Pa)
Butyl rubber	1.0	10^6
Metridium senile mesogloea	1.2	10^5
Pig thoracic aorta (tunica media + intima)	0.98	$\sim 5 \times 10^5$
Rabbit skin	20	10^8
Rhodnius, Triatoma tergal cuticle	1.4	2.5×10^8
Calliphora larval cuticle	1.8	10^7
Locusta intersegmental membrane	0.6–0.2	10^3

From Piez and Gross (1959); P. P. Purslow (personal communication)

work is (force × distance) then a soft material may require more work to fracture it than a stiffer one since it can stretch that much further under the same force (Fig. 4.20). However it is almost certainly also due to the strange J-shaped stress--strain curve of soft biological tissues (Gordon 1978). Ideas on the way this curve is generated have been mentioned above, but the significance may largely be in the control of fracture properties.

The fracture toughness of skin, which is difficult to tear, is of the order of 10–20 kJ m^{-2}, an order of magnitude less than that of aluminium foil, which tears very easily. However, it is important to decide just what we mean when we say something is more difficult to tear. Is it the ability to take higher loads for a given fracture toughness? Or greater extensibility at the same toughness? If one considers only the site of fracture, then it seems that it doesn't much matter what shape the stress–strain curve is (Kendall & Fuller 1987). But only teeth or the knife can produce such effects; a blade (as in a microtome) will always use the lowest possible work to fracture largely because of this. Atkins & Mai (1989) in their riposte to Kendall & Fuller come to the conclusion that biological materials are resistant to rupture because of their low modulus up to fairly high strains. This was largely substantiated by Purslow (1989) who

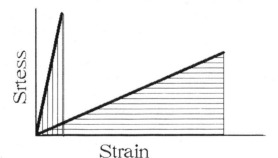

Figure 4.20 The stiffer material is less tough since it takes less energy (area under the curve) to break it. It is, however, stronger since it reaches a higher stress.

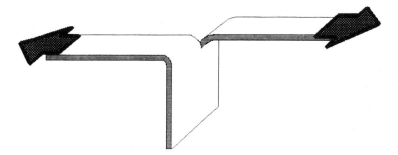

Figure 4.21 Morphology of the trouser tear test.

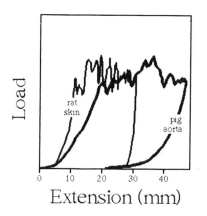

Figure 4.22 Force deflection curve of materials torn in a trouser tear test (Purslow 1989).

considered (as the other combatants had done) the trouser tear test (Fig. 4.21) which produces force-deflection curves of a fairly consistent type (Fig. 4.22). In such a test the work to fracture can be calculated simply by measuring the area enclosed by the graph (which represents the work used for fracture) and dividing it by the area cleaved. In this respect, the test is very much like the type of test discussed at the end of Section 1.6 in its usefulness with biological tissues. However, Purslow maintained that it didn't much matter what shape the stress–strain curve was, so long as there was sufficient material remote from the fracture zone where strain energy could be stored. In fact he and some Italian friends had earlier shown (Purslow *et al.* 1984) that the reorientation of fibres is probably the most significant factor. He proposed that when a crack or notch is introduced into a sheet of soft fibrous material a pattern of stress concentration is set up around the notch which determines the redistribution of the fibres. Using large-angle X-ray diffraction he measured the degree of orientation of collagen fibres around a slit made in a sheet of aortic media (Fig. 4.23) which was being stretched in two directions at once. At the tip of the crack, the collagen

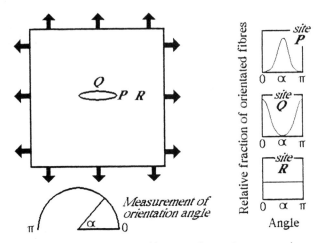

Figure 4.23 Orientation of collagen fibres at three sites around a cut in biaxially strained aortic tunica media (after Purslow *et al.* 1984).

is orientated across the path of the crack effectively stopping it. This reorientation is very local, so that if the media is strained to 0.3, the reorientation has disappeared at a distance of only 0.5 mm from the tip of a 5 mm crack. Since the reorientation is introduced as a direct result of the damage, it follows that it constitutes an automatic response of the material to wounding. Exactly the same response has been observed, qualitatively, by Broom (1984a,b) who has notched thin sheets of cartilage cut from the femoral condyle at the hip joint of a cow and stretched them. Illuminated with polarized light in transmission, the reorientation of collagen at the tip of the notch is very easily seen, as are the strands of of collagen pulled from the torn surface. So it appears that the reorientation effect is available in any soft collagenous composite.

It is also interesting that artery and mesogloea have the same toughness, as measured in the trouser-tear test. This illustrates both the conflicting requirements of a tissue and the relevance of various parameters to the actual functioning and design of a biological material. One way of thinking about this might be to say 'Any biological material is subject to certain forces. Some of these are going to be important and will play a large part in the selection pressures of evolution of the material. Some of these are going to be less important and will not be selected for as rigorously. How much selective advantage does a particular change in the mechanics of a particular tissue confer on the organism? For the one-in-a-thousand chance that an increase of stiffness will allow me to survive longer, is it worth the extra energy involved in modifying the material?'

For instance, it seems highly likely that the toughness of mesogloea comes 'for free', probably as a function of reorientation of collagen as a response to damage. The anemone is concerned mainly with stiffness and time constants of the mesogloea and it is unlikely that the benefits of the toughness of its mesogloea will be apparent to *Metridium*. The same is true of the insect cuticles, although the locust intersegmental membrane is sometimes found to have torn and been repaired, so it may not be tough (or strong?) enough. But it has to be soft enough for its particular function and presumably this precludes its being tough. Since this particular material is used only seven or eight times in the life of the locust it can afford to design it close to the working limits and put up with a small number of failures, which are not lethal anyway. By contrast it is possible that the artery is sacrificing toughness for resilience. The factors which will make a tissue tough will increase the energy absorption and will detract from the elastic performance by emphasizing viscous deformation processes. But part of the function of the artery, especially the aorta, is to store elastically some of the energy of the pulses of pressure in the blood as it is pumped by the heart. The energy is fed back into maintaining the flow of blood between pulses and so tends to damp out the pulses. This makes the flow rate more uniform without reducing the velocity of the blood as might happen if the aorta wall were to absorb the pressure pulses and not feed the energy back into the propulsion of the blood. The aorta is rendered as safe as possible under these restrictions by, as mentioned above, being tougher in one direction than the other. In fact it is not possible to measure the toughness of an aorta in the longitudinal direction using a tear test since the crack tip turns until it is travelling circumferentially. This may be because the elastin and collagen are both fibrous but are more or less separated laterally, thwarting stress concentration (Section 5.23).

Skin is a very tough material, for whatever reason. It is logical, however, that the outer covering of an animal should be tougher and it also happens that skin in some parts of the body (over the belly for instance) is less tough than the skin elsewhere (e.g. the back). Skin is a complex layered material, made up of four laminae:

1. Epidermis (the surface layer, 'keratinized' see also Section 2.3.1).
2. Sub-epidermal connective tissue (the 'grain' layer from which leather is made).
3. Corium (dense collagenous tissue).
4. Hypodermis (looser connective tissue).

The skin or dermis, usually thought of as the mechanically important part of this structure, is made of layers 2 and 3, which is essentially a more or less random feltwork of collagen and elastin fibres embedded in a protein-polysaccharide (mostly hyaluronic acid and dermatan sulphate) matrix. Its mechanical properties vary over the surface of the animal and are affected by such things are pH and cross-linking agents such as formaldehyde, which affect the interactions between the components. Various diseases exist which are due to failure of the cross-linking of collagen and elastin. One such is lathyrism, which can be induced by eating the seeds of the sweet pea, *Lathyrus odoratus*, or its active principle, β-aminopropionitrile. Such diseases have been extensively studied biochemically but hardly touched from the mechanical point of view. They generally reduce the tensile strength of the tissue by a factor of 2–5 and

Table 4.6 Changes in stiffness and strength of the wall of the Graafian follicle as ovulation approaches

	Stiffness	Strength
Unstimulated follicle wall	1 MPa	8 kPa
Stimulated (prerupture)	0.1 MPa	1.5 kPa

From Rondell (1970)

increase the extensibility up to 15-fold. A rather similar change occurs naturally in the Graafian follicle (Espey 1967, Rondell 1970). Just before ovulation the follicle swells and ruptures. Approximate values for the mechanical properties of the follicle wall are given in Table 4.6. The stimulation to swell and rupture is probably controlled by progesterone. Extensibility increases by about six-fold at the same time and the follicle becomes more pliant and relaxes more quickly. As might be expected this results also in more permanent deformation. Histological evidence shows that the collagen fibrils in a stimulated prerupture follicle wall are 'dissociating', which could be a result of breakdown of the matrix, of links between collagen fibres or some change similar to lathyrism. The fibrils are fewer and smaller in the wall of a prerupture follicle. These mechanical changes can be induced *in vitro* by luteinising hormone, cAMP or progesterone, all at pharmacological concentrations. The suggestion is that these chemicals all stimulate the production of an enzyme from follicular tissue which breaks down the structure of the follicle wall. The most likely action of the enzyme seems to be reduction of the modulus of the matrix, but the evidence is only circumstantial.

These changes in the follicle take several hours. In the intervertebral ligament of the brittlestar similar changes can take place which allow it to rupture little more than one and a half seconds after the animal is disturbed, autotomizing an arm as the result (Wilkie 1978). A number of factors can be shown to affect the creep rate and strength of the ligament: K^+ increases both, its action being blocked by anaesthetics which may indicate that nervous mediation is involved. But pH and reduction in Ca^{2+} also affect creep and strength and the hypothesis proposed is that Ca^{2+} is removed from the matrix of the ligament by a glycoprotein in neurosecretory cells adjacent to the ligaments. However, the response *in vitro* is at least two orders of magnitude slower than it is *in vivo* so presumably some other factor is involved. In this example the change is that of softening of the tendon matrix, but the catch apparatus of the echinoid spine seems to be due to an increase in the viscosity of the matrix of the collagenous ligament which encircles the base of the spine. This stiffens the ligament. Holothurians such as *Holothuria scabra* increase the stiffness of their skin when they are disturbed (Motakawa 1988). This, too, is a collagen-based material. The stiffness increases by an order of magnitude, a response which can be elicited from isolated pieces of skin by mechanical stimulation. This effect can be mimicked by varying the ionic environment of the skin (Eylers & Greenberg 1989). The modulus of the skin in $CaCl_2$ or distilled water is about

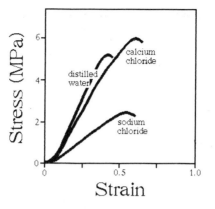

Figure 4.24 Stiffness of holothurian skin affected by various ionic environments (Eylers & Greenberg 1989).

18 MPa, but only 6 MPa in NaCl. The strengths also vary but strain at break is more or less constant at about 0.4 (Fig. 4.24). The hypothesis is that stiffness is regulated by cation-sensitive cross-links. The echinoid spine ligament will stiffen when exposed to 10^{-4} M acetylcholine, 10^{-6} M 5-HT or excess K^+, and will soften when treated with 10^{-6} M adrenalin. In the starfish the stiffness of the body wall of an animal which has been handled roughly is between 250 and 350 MPa, depending on orientation. But the stiffness falls off when the body wall is allowed to relax quietly (O'Neill 1989). So, as in the brittlestar, the change may be mediated by the nervous system.

4.7 STIFFNESS – A BIOLOGICAL VARIABLE

One way of describing the action of a muscle is to say that the force of contraction is due to an increase in stiffness. Consider a spring of low stiffness held at a constant strain: to hold a stiffer spring at the same strain would require a greater force or, alternatively, the force which is extending the weaker spring would not extend the stiffer spring so much. So if the pliant spring were stiffened but the force holding it extended not increased, the spring would contract. If echinoderms can change the stiffness of their ligaments over very short times it seems possible that they could use them for movement. A mechanism with some similarities occurs in the spasmoneme of the contractile stalk of protozoans such as *Zoothamnion* and *Carchesium*. The spasmoneme contracts more quickly and with more power than could an equivalent muscle thus pulling the main 'head' of the organism down onto the substrate – a defence reaction. Although the contraction was first thought to be due to a rubbery mechanism (Weis-Fogh & Amos 1972) it now appears that the spasmoneme proteins are binding Ca^{2+} in a very specific manner: with bound Ca^{2+} they tend to a more contracted configuration. The rather convenient thing about this is that once the spasmoneme is contracted no further energy input is required to keep it contracted, so the function is rather like that of postural muscle, although the energy saving seems not to have been calculated.

4.7.1 Prestress – rigidity from water

As will be seen in the next chapter, to make a rigid material for use in skeletal and supportive structures it is common to cross-link the components extensively and/or introduce large amounts of other components such as polyphenols or calcium salts. It is possible to make semi-rigid materials without resorting to this sort of ploy, using fibres and matrix in such a way as to form a prestressed structure. This is done by stabilizing water (usually in a gel) and surrounding and permeating it with a tensile fibrous web. In animals this form of structure occurs in cartilage (which forms articulating surfaces between bones and also pliant skeletal elements such as those of the nose, ear and intervertebral disc), and in plants it is responsible for the rigidity of unlignified tissues (the classical example is the flowering stem of the dandelion – and see Section 5.6 on cellular materials). In cartilage the fibrous phase is collagen (plus some elastin) and the matrix is polysaccharide (chondroitins, keratan sulphate and hyaluronic acid – glycosaminoglycans) stabilized by proteins in the manner of the lampbrush model of Fig. 4.2. This matrix can bind large amounts of water; the rigidity of the material (or structure) is derived from the osmotic swelling of the matrix against the restraint of the collagen fibres which are put into tension so giving a prestressed system, very similar in principle to a turgid plant cell. Just as the turgidity of a plant cell can be altered by putting the cell into a hyperosmotic solution which withdraws water from the cell, so if the osmotic relationships within the cartilage are affected by strong monovalent (or less strong di- or tri-valent) cations then the rigidity of the cartilage falls. Normally the content of water in cartilage is 65–80%, of collagen 15–25% and of glycosaminoglycan-protein (proteoglycan) 1.5–10%. So at the lower end of concentration of solids this material is not far removed from *Metridium* mesogloea, but there is more collagen and, of course, the arrangement and chemistry of the components are different.

The water-binding of the glycosaminoglycans is a function of the density of fixed charges just as in gels. This changes through the thickness of the cartilage (Fig. 4.25) so that the rigidity of the material probably also changes in a similar manner. In the end plate cartilage of a knee joint, as in the cartilage forming the end plate of the intervertebral disc system, the collagen in the superficial layers runs mostly parallel to the articular surface and in a preferred orientation; deeper into the cartilage the collagen is mostly perpendicular to the surface, and at an intermediate orientation in between (Fig. 4.26). In the intervertebral disc itself the collagen is orientated in a helical fashion, much like that observed

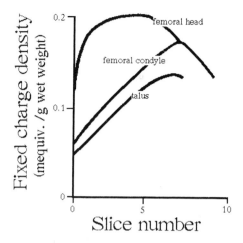

Figure 4.25 Change in the density of fixed charge with distance from the articular surface for cartilage taken from different joints (Maroudas 1975).

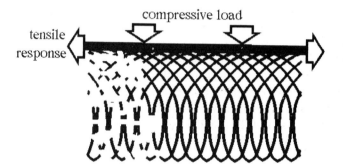

Figure 4.26 Orientation of collagen bundles in cartilage. The orientations actually observed are shown to the left of the figure; interpolation leads to the model to the right.

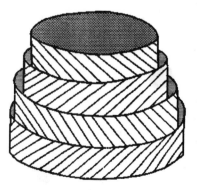

Figure 4.27 Diagram of nested layers of annulus fibrosus from an intervertebral disc showing orientation of collagen fibres (Hickey & Hukins 1980).

in worms and other fibre-wound pressure vessels such as rockets, gun barrels and cuttlefish (Fig. 4.27; Hickey & Hukins 1980). This will allow the layers of the annulus to deform much more readily (Wainwright 1988). Most mechanical testing of cartilage has been on human material from the larger joints, and most has been compressive. Under these conditions the cartilage shows stress–relaxation, partly due to normal polymeric relaxations but partly also due to the expulsion of water from the matrix. Stiffness is correlated with the proportions of proteoglycan present. Tensile tests on specimens from femoral condyles show that, as might be expected, the stiffness of the cartilage is also dependent on the content and orientation of the collagen. The impact properties of cartilage are also important – stresses at the bone–cartilage junction can be up to 8 MPa just in normal walking. An important function of cartilage is therefore that of resisting shocks, and the relative hydration of the matrix and its propensity to lose water under load will be important in this function. The cartilage is probably behaving elastically under these conditions – the shocks are probably mostly absorbed by the muscular system. There is also some evidence that the water content of the matrix is indirectly regulated by mechanically induced electric fields generated during walking and running. In the intervertebral disc system, the end plate differs from the sort found in other joints such as the knee in that it has a tangential layer of coarse collagen bundles arranged in sheets which connect the central area of the cartilage with the nucleus pulposus. The nucleus pulposus provides a stiff core to the intervertebral disc about which the two vertebrae rotate (Fig. 4.28). It generates this stiffness by having a gradient of fixed charges built into the glycosaminoglycans as uronic acid and hexosamine (Fig. 4.29; Urban & Maroudas 1979). These charged residues cause water to be imbibed into the disc, a process which is balanced

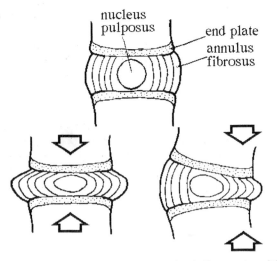

Figure 4.28 Change in shape of an intervertebral disc under different loadings (displacements are exaggerated).

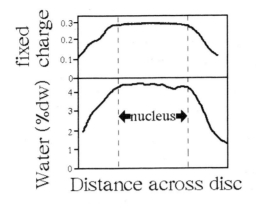

Figure 4.29 Change in the density of fixed charge (upper graph) and water content (lower graph) across a young intervertebral disc (Urban & Maroudas 1979).

by compressive forces on the disc from the weight of the body, from restraining muscles, tendons and ligaments, and from the restraint due to the collagen annulae of the disc itself. The molal fixed charge densities are about 0.3–mequiv/g H_2O which is equivalent to a swelling pressure of about 2 atmospheres. This is in agreement with the sort of load which has been measured *in vivo* in the nucleus pulposus of resting subjects.

Table 4.7 The stiffness of cartilage

Hyaline cartilage	25 MPa	(tension)
	500 MPa	(compression)
Costal cartilage	16 MPa	(tension and compression)
	Stress-relaxation	
	instantaneous E	relaxed E
Articular cartilage	2.3 MPa	0.7 MPa
Costal cartilage	7.8 MPa	5 MPa

From Wainright *et al.* (1965)

Figures for the stiffness of cartilage are few and various (Table 4.7). Most cartilage has a linear stress–strain curve, but cartilage from the ear has a concave stress–strain curve suggesting that the collagen fibres in this material undergo greater changes in orientation during straining. Such a concave shape could also be due to other factors, but illustrates a general point that skeletal materials tend to be Hookean.

CHAPTER 5

Stiff Materials – Fibrous Composites

In Chapter 4 all the materials considered are more or less pliant, consisting of stiff fibres in a pliant matrix. Under such circumstances the fibres act to introduce anisotropy (as in sea anemone mesogloea, locust extensible i.s.m., artery) to take loads due to internal pressures (cartilage, plant cells) or to take loads along their length (tendons). These materials are unable to resist bending or compressive loads since the fibres bend and move past each other. In order to make a material which can resist such loads it is necessary to increase the shear stiffness of the matrix. This will not only stiffen the matrix but also allow the transmission of stresses from one fibre to the next. Alternatively a new phase – ceramic material, usually calcium carbonate or phosphate – is introduced.

Such stiff materials are useful for a number of reasons. They can provide protection (mollusc and barnacle shells, the skull, sea urchins) shape and support (diatoms, arthropod exoskeleton) jointed limbs (arthropods, vertebrates) weapons of various sorts (teeth, tusks) and the like. All these materials are subjected to compressive, bending and shearing loads. This chapter explores the way in which fibrous composites, typified by insect cuticle and plant materials, are designed to cope with such mechanical demands; Chapter 6 is concerned with ceramics.

5.1 DOES STIFF ALWAYS MEAN BRITTLE?

The main problem about increasing the stiffness of a material is that the transmission of forces from one part to another is then that much better; a stiff material resists applied forces and does not stretch or flow to any extent. Whilst this means that it is possible to know precisely where different bits of a skeleton are relative to each other and thus to perform delicate and sequential movements even in the dark, relying solely on proprioceptive feedback, it does mean that a small mechanical defect can be introduced more easily into the stiff material comprising the skeleton. For instance you can scratch (and

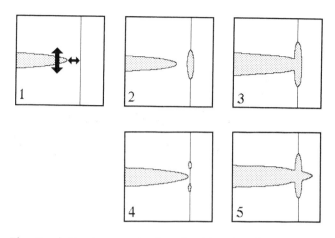

Figure 5.1 The Cook-Gordon mechanism for blunting the tip of a crack as it approaches a weak interface. Arrows denote direction and intensity of stress (Cook & Gordon 1964).

potentially weaken – Section 1.6) bone or stiff insect cuticle with a pin and leave a mark; you cannot do that with skin because it yields beneath the pin point. Because the stiff material does not flow so much under load, the defect serves to concentrate the forces onto relatively few chemical bonds and so bring them up to failure stress, even though the overall stress within the material may be much lower than this. Brittle fracture due to such stress concentrations is a well-known phenomenon and is used every day by the glazier when he scores, and then breaks, very easily and accurately, a sheet of glass. But if the tip of the crack can be rounded, the stress can be spread far more widely over many more bonds at the crack tip. This sort of thing can occur in materials which will tolerate high strains (not necessarily in all pliant materials – table jelly is pliant but brittle). In stiff composites the crack tip can be rounded at the interface between fibre and matrix: the mechanism for such an effect (Cook & Gordon 1964) is fairly widely known and accepted. It relies (Fig. 5.1) on the fact that the stresses at an advancing crack tip are not all in the direction tending to open the crack tip. A stress exists in advance of the crack tip and in the same plane as the crack, which is about one fifth the magnitude of the stress tending to open the crack. In a composite which has zones of weakness (fibre–matrix interfaces; zones of more pliant and/or weaker material) suitably arranged, this smaller stress will tend to open up a crack across the path of the main crack and so blunt its tip. The secondary crack can also absorb a large part of the strain energy driving the primary crack and so reduce the stress available for crack propagation.

Another way of avoiding the problems of brittleness in a stiff material is to protect the surface from defects. Such protection is probably the function of the sericin layer covering the fibrin core of silks. A similar sort of function is served by the protein matrix surrounding the chitin fibres of insect cuticle; the protein is probably protecting the fibres just as much as it is transmitting stresses to them.

5.2 SIMPLE COMPOSITE THEORY

5.2.1 Locust tendon

An analysis of insect cuticle as a classical fibrous composite has been made by Ker (1977) from which much of the following data are taken. Ker investigated the apodemes from the hind leg of the locust, *Schistocerca gregaria*, which are particularly convenient for a study of this sort. They are designed for tensile straining and are therefore appropriate for tensile testing; the average direction of the chitin microfibrils is parallel to the long axis with very little variation; the apodemes are highly uniform over relatively long distances; there is considerable information on their structure. It is therefore possible to analyse the mechanical properties of such tendons using the theory established for fibrous composite materials. We then need to ask how far does the theory fit in with the observed results and how do the results apply to other types of insect cuticle.

First the detailed morphology of the material: calculations relating electron micrographs of tendon to the chemical composition show that the microfibrils observed in cuticle are indeed chitin, that the matrix is protein, and that the two phases are separated. The chitin is strongly H-bonded (Fig. 3.6), so much so that it is insoluble in boiling 1 M NaOH. There are several models for the way in which the proteins are bonded to the chitin, but the most reasonable harks back to early work on chitin (Fraenkel & Rudall 1947) where the chitin was observed to have OH groups arranged with the same periodicity as the amino acids in a β-sheet conformation. The proteins are believed to have β-sheet areas which interact with and effectively adhere to the chitin. The bulk of these proteins is globular, as are the proteins forming the matrix in between the protein-covered fibres. Thus the matrix is very well bonded to the chitin fibres. This conclusion is supported by mechanical data: when the locust tendon is put under load there seems to be no slip at the fibre–matrix interface; distortion of the matrix occurs instead. Evidence for this is provided by the brittle failure of the preferred cuticle pulled longitudinally (Fig. 5.2) which shows that the fracture surface mostly goes straight across the sample with little or no evidence of the fibres being pulled from the matrix. So as the material is being loaded, if one fibre fails and slip does not occur at the fibre–matrix interface (which is the mechanism leading to fibre pull-out) stress will be concentrated near the broken ends and further failure at the cross-section containing the original break is made more likely. This is analogous to the method used by the glazier to break glass. There is, however, some evidence of a Cook-Gordon mechanism at work since there are several longitudinal cracks very much like the sort that occur in wood.

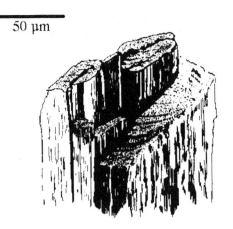

50 µm

Figure 5.2   The broken end of a locust tendon. Although it is substantially a brittle fracture, the top of the crack has been deflected by secondary cracks running vertically. This is very similar to the situation in wood.

. The longitudinal Young's modulus of the locust tendon is 11 GPa, the transverse Young's modulus is 0.15 GPa. The volume occupied by the chitin in the whole material (the volume fraction of the chitin) was estimated as 17%. The difference between the two moduli is due to the extreme degree of orientation of the chitin fibrils. When the tendon is pulled along the fibrils, both matrix and fibre are strained to the same extent. This is the equal strain (Voigt) limiting case for the behaviour of a composite (Fig. 5.3A). When the tendons are pulled across the fibrils both fibrils and matrix will be stressed to the same extent but will have unequal strains (the Reuss limiting case, Fig. 5.3B). The modulus of the tendon in this direction – to which the fibrils are not directly contributing – is lower, showing that the modulus of the matrix is lower than that of the fibres. The behaviour of fibre-reinforced composites has been modelled mathematically (Kelly, 1973) giving as the modulus of a composite like the locust tendon:

$$E_c = E_f V_f(z) + E_m (1 - V_f) \qquad \text{[Eq.5.1]}$$

where E_C is the stiffness of the composite measured along the fibre direction, E_f and E_m are the stiffnesses of the fibre and matrix respectively and V_f is the volume fraction of fibre. The quantity (z) is a complicated function which includes the stiffness, cross-sectional area, radius, spacing and length of the fibres and the change in shear in the matrix caused by the presence of the fibre. Using this model, Ker (1977) deduced that stiff arthropod cuticle, as exemplified by the locust tendon, conforms to the behaviour of a fibrous composite material yielding the data shown in Table 5.1.

The estimate of the stiffness of chitin given in Table 5.1 agrees well with that for cellulose and is the highest estimate yet made. The estimate for the stiffness of the protein matrix may be typical, but the stiffness of the matrix can vary

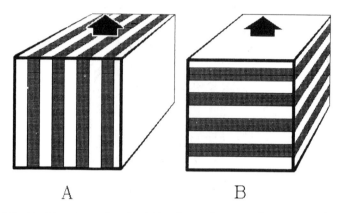

A   B

Figure 5.3 Limiting cases for the behaviour of composites: a, equal strain (Voigt) model; b, equal stress (Reuss) model.

greatly (see below). The estimate of the length of chitin fibril is interesting since it agrees closely with that obtained from the cuticle of *Sarcophaga* (larva, puparium and adult) using gel filtration techniques (Strout *et al.* 1976).

The fact that the chitin fibres are not continuous adds further complications to the factors involved in designing cuticle. The mathematical models assume that forces are not transmitted to the fibres across their ends. The stress must be transmitted by shear forces along the length of the fibres. Mathematical analysis and experiment show that the stresses are fed into the fibre from the matrix over a length – the transfer length – of the fibre. Thus the shear stress (τ) at the fibre–matrix interface has a maximum at the ends of the fibre and a minimum at the centre; conversely the tensile stress in the fibre (σ) is a minimum at the ends and maximum in the middle of the fibre (Fig. 5.4).

If, as seems very likely, this is true for chitin fibres in cuticle, this means that only infinitely long fibres can be strained to the strain of the composite. Shorter fibres are going to be strained by a lesser amount depending on the shear strain which is developed in the matrix as it transmits the stress to the fibre. This shear strain will be measured as part of the overall strain of the composite and represents the difference between the total strain and the strain on the fibres. Ker estimates that in locust tendon a quarter of the total strain is due to shear in the matrix. This sort of deformation will contribute to creep and energy losses if the bonds are relatively labile (i.e. H-bonds rather than

Table 5.1 Mechanical data for the components of insect cuticle

Stiffness of the chitin fibrils	70–90 GPa
Stiffness of the protein matrix	120 MPa
Length of chitin fibrils	0.36 μm
Degree of polymerization of chitin	about 700 residues

From Kerr (1977)

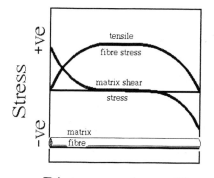

Distance along fibre

Figure 5.4 The variation along a fibre of the tensile stress, σ, and shear stress at the interface, τ, in a matrix–fibre composite.

covalent bonds), which seems to be the case with locust tendon. So it appears that the stress–strain properties of fibre and matrix, separately, could be represented as in Fig. 5.5. This figure shows the matrix to be elastic up to a yield point. The behaviour after the yield point may be viscoelastic or plastic. Whether or not the fibres break or flow with the matrix depends on the shear stiffness of the matrix and the efficiency of transfer of the stresses to the fibre. The fibre will not break if the matrix cannot transmit sufficient stress to it, either because the matrix is too pliant or because the fibre is too short. Under these circumstances the fibre is unlikely to contribute much to the stiffness of the material and the stiffness observed will be mainly that of the matrix (at small strains). One other factor which is important if the fibre is to stiffen the matrix

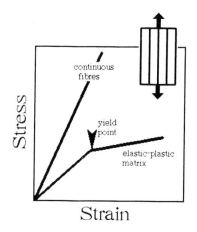

Strain

Figure 5.5 Diagrammatic stress–strain curves of continuous fibres and matrix of a composite loaded in tension (inset).

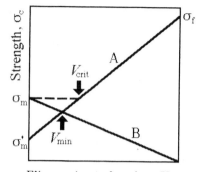

Figure 5.6 Theoretical relationship between the volume fraction of fibre and the strength of the composite. V_{crit} is the volume fraction of fibre necessary for the composite to be stronger than the matrix alone; V_{min} is the volume fraction of fibre necessary if the fibre is to reinforce the matrix effectively. A, $\sigma_c = \sigma_f V_f + \sigma'_m(l = V_f)$; B, $\sigma_c = \sigma_m(l = V_f)$

(still assuming that the fibres are all orientated along the direction of extension) is that:

$$\sigma_c = \sigma_f V_f + \sigma'_m(l - V_f) \geq \sigma_m \qquad [Eq.5.2]$$

where is breaking stress, subscripts c, f and m are for composite, fibre and matrix respectively and σ_m is the stress in the matrix when the fibres fracture. V_{crit} (Fig. 5.6) can be rewritten as

$$V_{crit} = \frac{\sigma_m - \sigma'_m}{\sigma_f - \sigma'_m} \qquad [Eq.5.3]$$

If the volume fraction of the fibres falls below V_{crit} then the composite will fail when the fibres break leaving only the matrix to take the load. The strength will then be given by

$$\sigma_c = \sigma_f V_f + \sigma'_m(1 - V_f) \geq \sigma_m(l - V_f) \qquad [Eq.5.4]$$

This gives a minimum volume fraction of fibres which must be exceeded if c is to be given by an equation analagous to Eq. 5.1, i.e.

$$\sigma_c = \sigma_f V_f + \sigma'_m(l - V_f) \qquad [Eq.5.5]$$

Equation 5.4 can be rewritten as

$$V_{min} = \frac{\sigma_m - \sigma'_m}{\sigma_f + \sigma_m - \sigma'_m} \qquad [Eq.5.6]$$

Taking, from Ker, values of 4 GPa for σ_f, 7.5 MPa for σ_m, and a value for σ_m' of $50 \times \sigma_m$ (a generous estimate), then V_{min} is about 8.5% and V_{crit} about 10%. Since the chitin content of the tendon is at least 17% according to Ker, then it is clear that it is the chitin which is taking the loads and not the matrix. In view of this Hepburn & Chandler (1976) were obviously wrong when they said, of stiff cuticles, that the matrix is the major contributor of stiffness and strength. They said this on the basis of simple tensile tests and had done no tests or calculations taking composite theory into account.

Very few cuticles have the chitin orientated in a single unique direction: amongst the stiff cuticles only those which are taking the forces along the chitin fibres (such as locust tendon) show such preferred orientation. Most cuticles have the chitin orientated in many directions (Fig. 5.7), mostly giving a plywood effect.

The orientation of chitin around the holes (pore canals, hair and bristle insertions) and other discontinuities varies in a rational manner to carry the stress trajectories around the obstacle and so avoid stress concentrations. The effect of changing the orientation of the chitin is to lower the modulus of the composite to ⅜ for the full plywood effect (as in Fig. 5.7a); the advantage gained is that the material is isotropic in the plane, as opposed to the extreme anisotropy of stiffness in the tendon which amounts to 530:1.

In such plywoods the fracture plane is deflected (Fig. 5.8) to give a very complex fracture surface much greater in area than that of a simple cross-section. Although the toughness of such a system has not been measured or compared with the fracture energy of something like the tendon which fractures much more cleanly across the chitin fibrils (Fig. 5.2), it is likely that a difference of

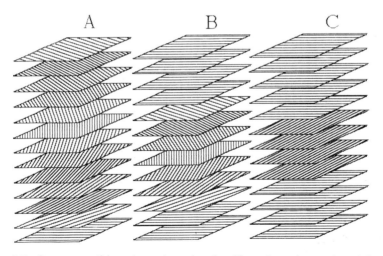

Figure 5.7 Some possible orientations for the fibres in arthropod cuticle. Each column represents a stack of layers with the orientation of the fibres in each layer sometimes the same as, sometimes different from, the orientation in adjacent layers.

Figure 5.8 Typical fracture produced from a cuticle with layers as in Fig. 5.7C.

at least an order of magnitude in the fracture energy is gained by the plywood type of orientation. Once again, a large part of the properties of biological 'materials' arises from the fact that they are actually structures (Section 5.6).

So far the mechanical behaviour of cuticle as a composite has been considered only in extension. In fact it is more likely that cuticle will be deformed in bending or compression, and so will be subjected to shear and compression. This mode of loading of composites has been hardly studied at all. An early study proposed that the material would fail by buckling of the fibres; the fibres can either buckle in phase (Fig. 5.9, buckling mode) when the matrix will fail in shear, or they

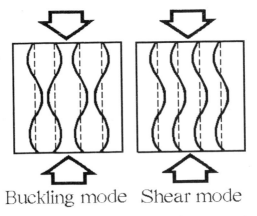

Buckling mode Shear mode

Figure 5.9 Possible behaviour of a composite under compresion along the fibres (Rosen 1964).

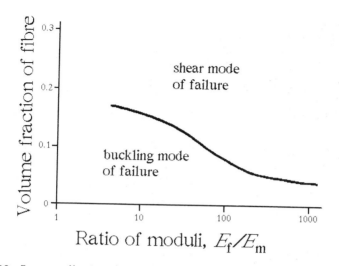

Figure 5.10 Factors affecting the moder of failure of a composite in compression (Schuerch NASA Report CR 202).

can buckle out of phase, when the matrix will be subjected alternately to compression and extension (Fig. 5.9, shear mode). However, calculations based on these models give answers considerably higher than those observed, which suggests that other factors are of importance. The type of failure (buckling in or out of phase) depends on the volume fraction of the fibres and the ratio between the modulus of the fibre and of the matrix (Fig. 5.10). Since nearly all insect cuticles have a chitin content of more than about 15%, and the only cuticles so far known which have lower chitin contents (e.g. that of the tick, *Boophilus*, which has about 3% chitin) are very pliant and so have a high E_f/E_m ratio, insect cuticles may be expected to fail in the shear mode only. This is rather convenient because the sum to calculate the critical stress is much simpler. If the shear deformation in the fibre is small compared with that in the matrix, then the critical compressive stress is given by:

$$\sigma_c = \frac{G_m}{1 - V_f} \qquad\qquad [\text{Eq.5.7}]$$

For locust tendon this becomes 48 MPa or about 7 g taking the area of the tendon into account. The tendon is not only unlikely to meet forces in this direction, it will probably bend before failing in compression.

5.2.2 Grass leaves

Plants, not much mentioned so far in this book, can also provide much amusement when analysed as composite materials. They are primarily made from cellulose which is present in the walls of the cells which in turn are commonly arranged in columns, which may also form fibres. So the mechanical properties of plants can be reduced to a combination of fibre composite theory (at whatever level in the hierarchy seems appropriate) and to the concepts of cellular materials (Section 5.6).

In many species of plants, notably monocotyledons such as grasses, the veins run parallel along the leaf so the leaf is essentially linear and in tension conforms to a simple Voigt model with the sclerenchyma bundles as the 'fibre' phase, contributing about 95% to the overall stiffness at a V_f of about 0.045 (Vincent 1982). Information about the ability of the grass leaf (and other materials with parallel fibres) can be gleaned by testing the strength of a test piece which has been damaged in a controlled manner, such as by the introduction of a notch or cut. The notch directs the fracture so that the test piece does not break at the clamps; depending on the material it may also cause a stress concentration. In a plot of tensile strength versus the relative length of the notch (expressed as a fraction of the total width of the specimen) a straight line (Fig. 5.11) indicates a 'notch-insensitive' material which can sustain damage without being greatly weakened. In other words, the strength of the material is a simple function of cross-sectional area. Grass leaves and the lamina of the seaweed *Laminaria* fall into this category (Vincent 1982, Vincent & Gravell 1986). If the line falls lower (Fig. 5.11) it indicates that the material is 'notch-sensitive' and can be weakened by the presence of small imperfections. This is potentially dangerous for the integrity of the material since a small crack may severely weaken the material (Section 1.6). Up to about 10% by volume of sclerenchyma, grass leaves are

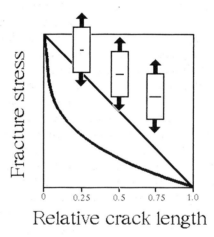

Figure 5.11 Sensitivity to notches or imperfections: in a 'notch-insensitive' material the notch merely reduces the strength in proportion to the amount of load-bearing material remaining (straight line). In a 'notch-sensitive' material a small imperfection of less than a tenth the width of the specimen can severely weaken the material (curved line).

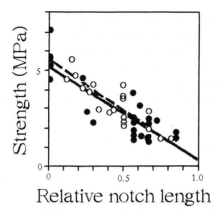

Figure 5.12 Effect of notches on a leaf of *Lolium perenne*. Open circles and dotted line = centre notch; solid circles and solid line = symmetrical edge notches. The lines are calculated regressions.

almost completely notch-insensitive (Fig. 5.12). This is most probably because the shear stiffness of the cells between the fibres is relatively low. Thus even if several fibres have been broken, stress is not sufficiently transmitted laterally to cause a stress concentration in the remaining fibre(s) and a notch will not weaken the leaf significantly. This is not so with *Stipa* which has a continuous sclerenchymatous layer over the upper surface of the leaf. This gives a V_f of sclerenchyma of about 0.3 (Fig. 5.13). A crack can propagate through this layer and the *Stipa* leaf breaks in a brittle manner (Fig. 5.14; Vincent 1990).

The notch-insensitivity of the softer grasses has a number of important consequences for animals feeding on it, since it means that teeth are of little use other than for gripping the grass. Like meat (and for the same reason) it must be torn by brute strength or cut with a shearing action. Large animals such as cows and sheep hold the grass (respectively with tongue and teeth) and pull. They thus have a limitation on the number of leaves which they can break, illustrated by considering the effects of 'enrichment' of pastures. Longer grasses will be harvested at least as easily as shorter ones, so it is worth encouraging grasses to grow tall. But if the grass tillers grow more densely the cow or sheep, having finite strength, will find the size of its bite reduced, even though it might be taking in the same number of grass leaves. This is because the strength of grass and its apparent stiffness (as measured from the total cross-section area of the leaf) are directly proportional to the amount of fibre present. Thus increasing tillering will increase intake only up to a point (Antuna, A., personal communication). A number of different studies have shown inverse correlation between the strength of grass and 'palatability' (e.g. Theron & Booysen 1966) so that the weaker grass is, the more the animal will eat. It has also been pointed out that when deer age their teeth tend to rot. If their back teeth remain, food intake is not impaired. However if their back teeth decay they cannot survive,

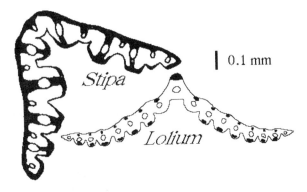

Figure 5.13 Sections through leaves of *Stipa gigantea* (left) and *Lolium perenne* (right). Sclerenchyma is shown as solid dark areas; the scale line is 0.1 mm.

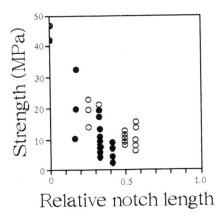

Figure 5.14 Notch sensitivity of *Stipa gigantea* showing weakening by small notches.

even if their front teeth are in good condition. Clearly deer do not need their front teeth for gathering grass, but need their back teeth for comminution. Smaller animals (rabbits, locusts etc) have to cut through the individual fibres of the leaf since they are not strong enough to break the grass in tension.

As grass dries the stiffness increases but the work to fracture hardly changes at all (Vincent 1983). At a water content of about 0.2 g per g dry weight^{-1} of grass, the fracture properties of the cells between the fibres go through a transition and these cells become very brittle and weak. The grass now tends much more to fracture between the fibres, and to fracture in a very brittle fashion. This seems to be the condition for hay-shatter; hay containing 25% or so of water is considered to be suitable for bringing in from the field. Hay with a lower water content is considered to be very susceptible to shatter with consequent high losses of the crop. Water content has been shown to be the major factor in hay-shatter in non-grass hay species (Shepherd 1961).

5.2.3 Horn keratin

Keratin has already been mentioned as being a fibrous composite (Section 2.3.1) which can be effectively modelled using composite theory (Kitchener & Vincent 1987). Horn provides a very convenient source of keratin in large pieces. Using the Voigt model and the experimental data shown in Table 5.2, the calculated stiffness is correct for totally dry horn but becomes progressively larger as the water content of the matrix increases. In the totally dry horn, the stiffness of the matrix and fibre are assumed to be equal, which then gives an estimate for the stiffness of the fibrous phase. This is reasonable, since as the matrix dries out, the stiffness of the matrix gradually approaches that of the fibre (Fraser & MacRae 1980). 'Fresh' represents the amount of water in the horn when it is freshly removed from the animal; 'wet' represents the largest amount of water which can soak into the matrix. The shear stiffness of the matrix is thereby reduced, increasing the stress transfer length (Fig. 5.3). Further calculation along the lines used by Ker on locust tendon showed that a fibre length of 40 nm gives a very good estimate of the stiffness of both fresh and wet horn (Table 5.2), but this is much shorter than reported lengths of protofibrils (the morphological equivalent of the fibres in the mechanical model) of 500 to 1000 nm measured from electron micrographs. The resolution of this problem may come from a model for wool in which the protofibrils have a 5 nm length of 'disordered' or amorphous material every 15 nm (Feughelman 1979). This is rather like the model for silk. These disordered regions are staggered with respect to each other. Perhaps a similar model could apply to horn keratin. Then when water is taken up by the matrix the disordered regions of the protofibrils will also be hydrated so that the integrity of the fibres will be affected.

The fracture properties of horn are very much like those of grass, in that fresh and 'wet' horn are both insensitive to notches (Fig. 5.15). This is obviously of great importance to an animal such as a billy goat whose legendary reproductive prowess is predicated by some pretty hard fighting, involving forces

Table 5.2

	Dry	Fresh	Wet
Water content (%)	0	20	40
Mean bending stiffness (GPa)	6.1	4.3	1.8
E_f (GPa)	6.1	6.1	6.1
E_m (GPa)	6.1	3.1	0.9
V_f	0.61	0.56	0.53
V_m	0.39	0.44	0.47
G_m (GPa)	2.3	1.1	0.9
estimates of bending stiffness			
Voigt (GPa), infinite fibre length	6.1	4.8	3.6
As above, fibre length of 40 nm	6.1	3.8	1.9

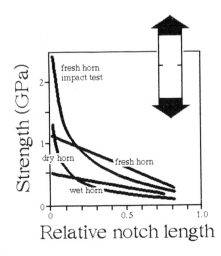

Figure 5.15 Notch sensitivity of horn keratin (Kitchener 1988).

up to 3.4 MN (Kitchener 1988). The wet horn protects the bone horn core from damage, and does so more effectively if it is damp or wet, presumably because the matrix will then yield and flow, dissipating any stress concentrations. This apparently explains, rather nicely, an aspect of the behaviour of these animals before fighting – they plunge their horns into mud or wave them around in a dewy bush. Students of animal behaviour called this 'horning' and put it in the 'flight or fight?' category of behaviour – the goat's equivalent of biting its nails. It seems more likely that the goat knows more about its horns than most of its observers and is toughening them up for the fray. With only one horn both his fighting and sexual activities will be curtailed.

5.3 TO STIFFEN THE MATRIX

The importance of the shear stiffness of the matrix has been emphasized. In grass this is apparently due to loss of water; in insect cuticle the origin of this stiffness is obscure and a subject of some dispute. The problem arises because the matrix is produced as an apparently amorphous protein in solution which has to be stiffened by the introduction of cross-links. The classical story, proposed by Pryor (1940), is that dihydroxyphenols (Fig. 5.16) are secreted into the cuticle and that these are oxidized to quinones within the cuticle. The quinones are bifunctional and cross-link adjacent protein chains. This scheme has been accepted, more or less unquestioningly. However, a paper was published very shortly after Pryor's, in the same journal, by Fraenkel and Rudall (1940). In this the hypothesis was advanced that the most dramatic event occurring in the cuticle when it tans is the loss of water and that this is the most

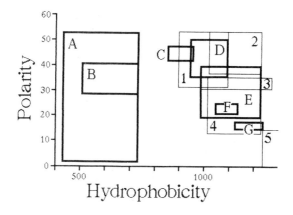

Figure 5.16 Hydrophobicity/polarity of amino acids in insect cuticles and a number of other proteins. A, insect silks; B, resilins; C, fly larval and puparial cuticles; D, termite cuticles; E, beetle cuticles; F, Extensible cuticles from *Rhodnius* and *Boophilus*; G, locust sclerite cuticle; 1, high-sulphate keratins; 2, globulins and albumens; 3, collagens; 4, high-tyrosine keratins; 5, elastins (Vincent 1980b).

important factor. This paper is often totally overlooked. A further factor is that Pryor worked with the egg case of the cockroach, whereas Fraenkel and Rudall worked with the cuticle of a fly larva. The cockroach egg case has a number of advantages as a model system: it contains no chitin, only protein, so that any stiffening must be due to cross-linking the protein and not chitin–chitin or chitin–protein links; it is produced from time to time throughout adult life from a pair of glands (the colleterial glands) whose contents can be analysed, allowing the precursors of the tanned protein matrix to be investigated; the egg case is tanned as it is produced, so the process can be followed in a single egg case. By contrast the cuticle of the maggot contains chitin; the processes going on within it must be inferred from what can be extracted from the cuticle, which is difficult if the processes being investigated are stabilizing the components and rendering them inaccessible. However, Pryor's model has one highly significant advantage. The phenols involved are fairly reactive, so elegant biochemistry was possible. By contrast, experiments on the control by a cell of the water content of its surroundings were, and are, difficult to perform. It is probable that a major factor in the almost universal acceptance of Pryor's model is that it is easy to manipulate. But the basic problem is that in all cuticles investigated the two processes of phenolic tanning and dehydration occur together: it's a classic case of having two variables (at least) in a single experiment. Can biomechanics supply the second experiment?

The basic observation is that the proteins of the cuticle matrix become cross-linked and stiff. This does not necessarily require covalent bonds. In fact the only cuticle in which covalent cross-linking has been shown is resilin (Section 2.3.4.2). This is inferred because the amount of cross-linker as calculated by rubber elasticity theory matches the amount estimated from biochemical experiments; the material is hardly viscoelastic showing that the protein chains are not interacting between the cross-linking sites (if they were, the making and breaking of secondary bonds would introduce a viscous component) and the material does not creep. In addition the protein of resilin has a low hydrophobicity index (Hillerton & Vincent, 1983) which suggests that there is very little interaction between the protein chains when the material is hydrated. Since resilin is a cuticular protein in which the peptide chains do not interact, it is possible to calculate how much stiffness is contributed to tanned cuticle by covalent cross-linking (if it actually occurs – assume for the moment that it does) on its own. The pre-alar arm (an element in the wing hinge) of the locust contains resilin and chitin and has a Young's modulus of 2 MPa. The density of cross-linking is equivalent to the addition of about 5% by weight of cross-linker. Tanned cuticles can have more than this added at sclerotization: the largest increment probably occurs in the fly puparium where there is an increase in dry weight of about 25%. If this were all cross-linker, and if this amount of cross-linker were added to a cuticle with a protein like resilin, the stiffness would be about 10 MPa (on rubber elasticity theory) which is certainly not as high as stiff cuticle and not even as stiff as the protein matrix of the locust tendon. Thus the introduction of phenolic cross-links, in the absence of any other sort of interaction between the proteins, cannot account for the stiffness of tanned cuticles. The proteins in such cuticles must be interacting far more extensively to reach such high stiffness, and such secondary interactions will be greatly encouraged and stabilized by the removal of water from the system. The removal of water will be aided by such factors as high hydrophobicity of the proteins, and the stiffer cuticles of the locust have a more hydrophobic matrix (Fig. 5.17). It must be emphasized that this is a rather rough-and-ready index and that it is here being applied to a mixture of proteins rather than to a single protein.

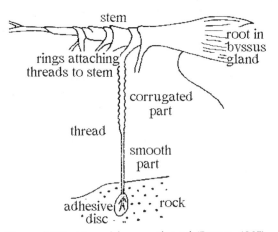

Figure 5.17 Mussel byssus thread (Brown 1965).

Table 5.3 Stiffness of fly larval cuticle under differing conditions (initial modulus – i.e. at small strain)

Source of cuticle	State of hydration	Orientation of chitin	State of tanning	Stiffness
3rd instar,	wet	helicoidal	none	25 MPa
fully fed,	wet	helicoidal	6 h in 3% catechol	35 MPa
ready to				
pupariate.	wet	helicoidal	3 days in 3% catechol	88 MPa
	dry	helicoidal	none	590 MPa
	dry	helicoidal	3 days in 3% catechol	590 MPa
White puparium	wet	preferred	none	73 MPa
	dry	preferred	none	2.2 GPa
Tanned puparium	wet	preferred	naturally	250 MPa
	dry	preferred	naturally	3.1 GPa

Whatever the mechanism involved, it looks as if water is important in controlling the matrix properties of insect cuticle just as it is important with collagens and keratins (Chapter 2). Certainly the water content of a cuticle can be shown to be closely correlated with the stiffness: the cuticle of the maggot can be tanned artificially in dihydroxyphenols such as catechol. Up to 20% of the dry weight of a cuticle tanned in this way can be catechol (Vincent & Hillerton 1979). Solubility of the proteins is reduced but the stiffness is only slightly affected if the cuticle is not allowed to dry (Table 5.3). Air-drying of larval cuticle, tanned or untanned, produces a material of much higher stiffness, so tanning on its own makes no real difference to the stiffness. The same is true of cuticle taken from maggots which are beginning to pupariate. The first stage in pupariation involves the shortening and rounding of the larva. At the same time the cuticle doubles in thickness and chitin fibrils are orientated circumferentially. The solubility of the proteins is unchanged, showing that tanning has not yet started. A piece of this cuticle, dried and tested in tension circumferentially has the same stiffness as a piece of dry tanned puparial cuticle (Table 5.3) and the response is Hookean, suggesting that the tension is being taken directly by chemical bonds. This would be unlikely if the impregnating phenols were the only factor responsible for giving stiffness since they are much more widely spaced.

Why add phenols? What is their purpose if they cannot to stiffen the cuticle by their cross-linking action? One of the original suggestions by Pryor and by Fraenkel and Rudall was that the phenols react with polar side-groups. This reaction displaces water molecules and so achieves dehydration. Pryor showed that a lump of gelatin in strong benzoquinone solution tans, goes darker, shrinks, dries and becomes leathery. So the phenols may be initiating or accelerating the dehydration. It is also possible that they are polymerizing to act as filler particles. The dark areas in cuticle which occur around joints and in hairs and mandibles are places where large amounts of melanin – an indol polymer – have been deposited. Professor Andersen of Copenhagen (Andersen 1986) has shown that other phenols will polymerize in the cuticle. So the phenols may be more important in filling the spaces between the protein chains and reducing the chance of voids in the matrix which could start cracks.

What is the nature of the 'sclerotized' matrix? Presumably this will depend on a number of factors. If the proteins are more hydrophobic, they will tend to be globular in an aqueous environment and there is reason for thinking this is so in locust tendon. This may lead to a phased structure with strong secondary bonds within the globules and cross-links of some sort between the globules. But not all tanned cuticles have a high hydrophobicity index. The cuticle of *Calliphora* larvae is an example. This should mean that it has a high water content and very little interaction between the protein chains. And so it probably has in the untanned state. But just like silk, which also has a low hydrophobicity index, the protein of *Calliphora* larval cuticle is capable of forming beta structures. If such cuticle is extracted into 7 M urea, the extract dialysed against water and the dialysed extract dried and examined in the infra-red it will be found to be largely β-sheet (Hackman & Goldberg 1979; Hillerton & Vincent 1979). The protein in this conformation is largely insoluble. So it is possible that, in some cuticles, dehydration causes the formation of crystalline structures in the matrix which then becomes as insoluble as chitin, and for much the same reasons. This mechanism was originally proposed by Fraenkel and Rudall (1947). In conclusion it might be pointed out that if the causes and mechanisms of sclerotisation are still uncertain it is probably due in part to the fact that there seems to be no single set of circumstances for producing a stiff cuticle. It is to be expected that if the aim is to introduce numbers of primary and/or secondary cross-links there will be many ways of doing so.

5.3.1 Other sclerotized systems

A number of other tanned protein systems exist. The way these are usually recognized is by the colour (typically brown), the stiffness and presence of phenols in special glands or cells. Quite often such systems are associated with chitin. The hard brown sheath around *Obelia* (Coelenterata, Hydrozoa) has been shown to be made of a protein-chitin material (Knight 1968): phenol precursor (catecholamine), phenoloxidase and protein are found in vacuoles in tanning cells which are especially numerous in growth areas. The egg shells of many turbellarians, trematodes, nematodes and some cestodes are tanned protein as are the chitin-containing hairs of many annelids, the tubes of *Sabellaria* and the like. An interesting tanned system is the silks which many moths use for their cocoons, especially when these are exposed to the elements. An example is the burnet moth (*Zygaena* sp.); *Antheraea pernyi* and *Samia cynthia* also produce tanned silks. Brunet & Coles (1974) measured the mechanical properties of these latter two silks and found that although the initial Hookean part of the curve was not changed, indicating that any phenolic tanning was not affecting

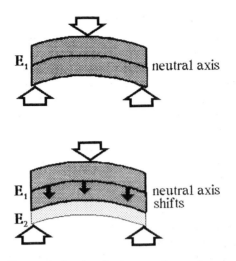

Figure 5.18 Shift in the neutral axis of a beam when a layer of more compliant material is added (lower figure). Energy is stored by flattening the beam, thus putting the added layer into tension (Ker 1977).

the stiffness, it did appear that the tanning affected the yield behaviour, making the silk much tougher. It seems likely that the tanning is waterproofing the silk (Vincent 1980b). The outer covering of the shells of many bivalves (the periostracum) such as *Mytilus* is made of tanned protein and may contain chitin: the same is true for many gastropods. The byssus threads of *Mytilus* (Fig. 5.18) are phenolically tanned, as, it seems, is the glue which holds them on to the rock. Again the analysis recognizes that water is an important component in the system. The prime requirements of a technological adhesive are that it should be able to join dissimilar materials which are both dirty and wet. Water is the biggest problem and adhesives technologists go to great lengths to exclude water, both during the creation and setting of the joint and during its useful lifetime. In a long series of papers, Herb Waite and his co-workers have analysed the adhesive technology of the mussel byssus (Waite 1987) to the extent that they can manipulate it using the techniques of biotechnology (Benedict 1987). Benedict's patent claims for the production of 3,4-DOPA from tyrosine which is then reacted with tyrosinase, thus forming a bioadhesive polyphenolic protein. This presupposes that all the problems which the unadhered mussel has to face have been solved. Here is Waite's version of how the mussel manages. Waite likens the end of the mussel's foot, which finally moulds the adhesive plaque at the end of the byssus thread, to a rubber sink plunger. It presses firmly onto the substrate and scrubs it clean, pressing the water away at the same time. The foot is then lifted away from the substrate leaving reduced pressure in the cavity beneath it, into which granules of precursor adhesive are secreted. The adhesive then develops into a foam structure – how or why is not known. The degree of adhesion to a number of types of surface of varying hydrophobicity is shown in

Table 5.4 Adhesion to surfaces of varying hydrophobicity

Surface	Strength of adhesion (MPa)
Teflon	0.015
Paraffin wax	0.015
Acetal	1.2
Glass	7.5
Slate	8.5

From Waite (1987)

Table 5.4. In general it seems that the adhesive has to compete with water for interactions with the substrate, so the more the water interacts, the more competitive the adhesive has to be. Analysis shows the adhesive to consist largely of a repeated sequence of ten amino acids: -Ala-Lys-Pro-Ser-Tyr-Hyp-Hyp-Thr-Dopa-Lys- or a subset of these. This allows for a number of different types of interaction with a variety of substrates, but the most interesting and unusual aspect is the presence of Dopa. It obviously means that the various interactions which stabilize insect cuticle, including interaction with metals, is possible. In fact, a number of industrial patents exist in which catechol (i.e. 3,4-dihydroxyphenol) derivatives are used as the basis of metal primers. So it's the ideal way to get a mussel sticking onto a metal post in the sea. The hydroxyl groups associated with serine, threonine and hydroxyproline also allow the decapeptide to compete for water binding sites, so the whole effect is of hydrophobicity and waterproofing. This could be emphasized by final cross-linking of the phenolic groups in the same way as proposed for insect cuticle and silks.

The mussel glue is under development for production as a commercial adhesive, to be used in wet and biologically sensitive situations (e.g. sticking tooth crowns in position). But blind copying, as is achieved down the biotechnology pathway, may not be the best route. Kaleem *et al.* (1987) pointed out that proteins are the most under-rated and under-used polymers. They have great mechanical strength and inherent adhesive properties – all that is necessary is to exclude water and enzymes which can digest them. This has proved to be possible using gelatin which was chemically modified by (guess what?) grafting phenolic groups onto the lysine side chains. These were then further modified to make a sort of epoxy resin system. Their adhesive properties were only a tenth as good as an advanced epoxy adhesive, but then these are the end product of over 40 years of research. So this is a promising line and may represent a way not only to develop the adhesive properties of proteins but also to harness the superior properties of silks and collagens.

5.4 LAYERED COMPOSITE – IMPROVE TOUGHNESS

One of the characteristics of insect cuticle is that its properties change through its thickness. The outer layer of the cuticle is stiffened (by whatever mechanism)

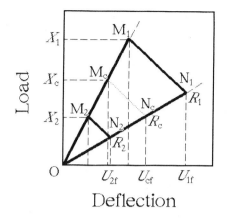

Figure 5.19 Load deflection curve illustrating the effect of sticking two layers of material together, then testing the layered composite as a double cantilever beam (Fig. 1.26). See text for explanation.

but the inner layer is usually unstiffened and relatively pliant. The nomenclature of these layers is confused since the process of stiffening the cuticle can continue long after the insect has ecdysed. Similarly the pliant inner layer often continues to grow. It is probably easier to call the stiffened layer the exocuticle and the more pliant layer the endocuticle. The easiest and most reliable way to differentiate these layers histologically is using Mallory trichrome: the exocuticle either does not stain and appears amber or stains red, and the endocuticle stains blue. In some cuticles there is a layer in between called the mesocuticle. The presence of these layers makes the interpretation of the mechanical properties of most cuticles very difficult, and several mechanical studies which have been performed on suchcomplex layered cuticles should be regarded with deep suspicion. However, such layered materials are also of interest to the materials scientist. Ker (1977) was interested in cuticle as an energy store in jumping beetles. The particular flea beetle he investigated uses the leg tendon rather like a coil spring being straightened. The energy is not then stored in simple extension as it might be in a catapult or as it is in the locust tendon, but stored in bending. Figure 5.19 shows a part of this tendon which can be considered as a beam. The upper layer represents the tanned cuticle and has a Young's modulus of E_1; the lower layer represents the untanned cuticle and has a Young's modulus of E_2. The bending of the beam is such as to compress the tanned cuticle and stretch the untanned cuticle, and is shown by the broken lines. Ker showed that this system could store 2.25 times more energy in the tanned cuticle if the untanned layer were a third the thickness of the tanned layer with $E_2 \ll E_1$. His overall conclusion was that adding an untanned layer is beneficial so long as this difference between the moduli stays at a factor of about two. The improvement in the amount of energy stored is due to a shift in the neutral axis (the line along which neither tension nor compression occurs) within the tanned cuticle. As an alternative to increasing the energy storage per unit volume, this shift of the neutral axis may be considered as allowing for a greater distortion at given volume of tanned cuticle. The cuticle thus not only provides a stiff outer covering but, owing to its layered structure of exo- and endo-cuticle, is designed to store or absorb energy. This greatly increases the ability of the cuticle to resist impact loads (such as a heavy boot with your food inside) without rupturing. The same problem has been approached in a slightly different fashion by Guild *et al.* (1978) who considered the fracture toughness of layered structures.

The test morphology used for this analysis is the double cantilever beam (Section 1.6; Figs 1.30, 1.31). Two identical test pieces, 1 and 2, with different resistances to cracking, R_1 and R_2, will follow force-displacement curves (Fig. 5.20) OM_1-N_1 and OM_2-N_2. If these two materials are glued together they will have a composite behaviour somewhere between the individual responses (OM_c-N_c). Experiments with such a system show that with a suitable glue the crack propagates in the two layers at the same time so that the cracking in the less tough layer is suppressed (until OM_cN_c) and cracking in the tougher layer occurs earlier (at OM_c-N_c) than in the absence of gluing. This suggests that energy is transferred through the glued joint from the less tough layer to the tougher, although the total work done must be the same in both cases, glued or unglued layers. If these values for cracking load are related, it is found that the cracking load for the composite (X_c) is a geometrical and not arithmetical mean of the cracking loads for the two layers (X_1 and X_2):

$$X_c = [\tfrac{1}{2}(X_1^2 + X_2^2)]^{1/2} \qquad\qquad [Eq.5.8]$$

This result has been confirmed by model experiments. The value of a geometrical mean will always be greater than that of the linear mean. Applied to cuticle, this fact means that cuticle can be made stiff and linearly elastic, and thus capable of bearing loads for long periods without relaxing significantly, without incurring all the associated difficulties of failure at small strains. The stiffnesses of exo- and endo-cuticles are unlikely to be the same since the level of cross-linking and hydration is so different; also the thicknesses of the two layers will be different and the ratio between the two thicknesses varies from place to place. So it is unlikely that the above model will apply in the simple form presented here. For two layers, total load:

$$X_T = X_2 \sqrt{\left(1 + \frac{E_1 t_1}{E_2 t_2}\right) 1 + \left(\frac{R_1 t_1}{R_2 t_2}\right)} \qquad\qquad [Eq.5.9]$$

In addition the load at which each layer cracks, X', is given by:

$$X_1' = \frac{a}{1+a} X_T; \ X_2' = \frac{1}{1+a} X_T \qquad\qquad [Eq.5.10a,b]$$

where $a = E_1 t_1 / E_2 t_2$

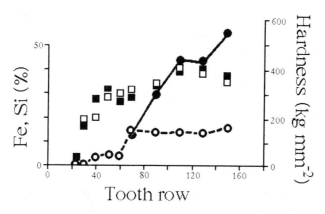

Figure 5.20 Variation in Vickers Diamond Hardness and mineral content in the teeth along the radula of *Patella vulgata*. Squares = percentage minerals (Fe as goethite, Si as opal) in the anterior (open squares) and posterior (filled squares) parts of the tooth. Circles = hardness in the anterior (open circles) and posterior (filled circles) parts of the tooth (data from Runham *et al.* 1969).

All this theory assumes Hookean elasticity of both layers. For a fibrous composite such as cuticle it will work only at very small strains. The apparent (incremental) modulus of the endocuticle will change with strain as the chitin fibres are orientated in the direction of strain; this may improve the toughness of the cuticle and be an added reason for the layered structure of cuticle. Certainly the exocuticle is more brittle than the endocuticle as is shown when cutting thin histological sections; the exocuticle will shatter, the cracks being stopped at the interface between exo- and endo-cuticles.

5.5   THE EFFECTS OF ADDING FILLERS TO COMPOSITES

One of the most interesting, most investigated and most characteristic molluscan structures, which is closely similar to insect cuticle, is the radula. This is, in its simple form as typified by the limpet *Patella*, a ribbon of tanned protein and chitin bearing teeth of much the same material. The work of Runham and his colleagues has failed to establish the nature of the tanning agent and has failed to detect a phenol oxidase (Runham 1961), although there is other circumstantial evidence for quinone tanning. However, iron and silica are incorporated into the radular teeth (Jones *et al.* 1935) and the teeth become harder during their development (Runham *et al.* 1969).

5.5.1 Hardness

The test used to measure hardness is an indentation test. For instance, the Vickers Diamond Hardness test involves pressing a pointed diamond, cut to a four-sided pyramid shape and of standard size, into the material under test. Since the diamond probe is tapered the size of the hole it leaves in the material under test is proportional to the depth to which it has been pressed, so simple measurement of the length of a diagonal of the hole gives a measure of the hardness. As the diamond pyramid sinks into the material it causes shear deformation. So the hardness measured will be closely related to shear stiffness. But the measurement which is made is the amount of deformation remaining after the deforming force is removed. So the shear strength of the material will also be important. After the deforming force has been removed the material is free to recover. So the ultimate shear strain of the material will also be important since that will influence the recoverable shear strain. The viscoelastic properties will also have an effect since the more viscoelastic the material is the longer it will go on recovering after the load is removed and the time at which the hardness measurement is made will be important. So hardness is really a relative term when applied to viscoelastic materials. It expressed in units of stress which is then equated directly with shear strength.

When it comes to putting a figure to hardness it is important to consider what this figure represents. A metal has a moderately homogeneous structure so hardness as measured by the indentation test gives a figure which, in most instances, can apply to the whole surface of the material. But in some metals the size of crystal is large and a different figure for hardness will be derived according to whether the indentation is made towards the centre of a crystal (which will be purer metal) or at crystal boundaries where additives to the metal will be present in higher concentration and the plastic behaviour of the metal will be different. So the structure of the material is important. This is the result of using a test which examines the material in a specific size range. A stress–strain test averages the response of the material between the clamps of the test rig; even so there is the possibility of end effects due to the clamps but this can be allowed for by using samples of different sizes. The equivalence for hardness would be a wear test of some sort, but this is by no means as convenient and takes much longer. Also the processes of wear are by no means well understood and rely on a number of factors which are difficult to control. The indentation test is convenient, but it must be used with insight. Coming back to the results for limpet radular teeth, Fig. 5.21 shows how the incorporation of iron and silica into the anterior and posterior parts of the tooth (Fig. 5.22) varies with the development of the tooth, and how the hardness of the tooth also varies. It is apparent that hardness as measured by the indentation test is not varying with the content of iron and silica (tentatively identified as opal) (Fig. 5.23). It certainly cannot be that due to the particles themselves: the Mohs scratch test which rates hardness from talc (point one on the scale) to diamond (10) places silica (as opal) at 5.5–6.5 and goethite (the form of iron reported for the radula) at 5–5.5 which is the opposite ranking to the hardness of those parts of the tooth containing these minerals. The iron and silica are present as particles [goethite particles are 0.2–1.0 μm long and 15–20 nm wide (Webb

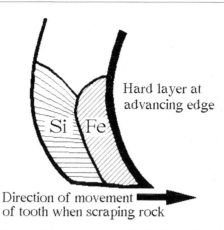

Figure 5.21 Diagrammatic section of a radular tooth from *P. vulgata*. Lines show the directions of the chitin fibres; Fe and Si are distributed as indicated.

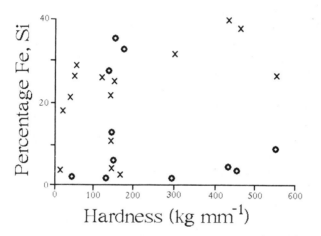

Figure 5.22 Hardness of the radular tooth of *P. vulgata* is not related to the amount of Si (o) or Fe (x) in the tooth (data from Runham *et al.* 1969).

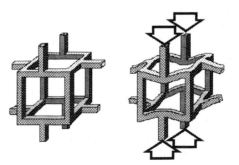

Figure 5.23 A cubic model for an open-cell foam (left) and the way the edges bend during linear elastic deformation (Gibson & Ashby 1988).

et al. 1989)] yet the indenter makes indentations about 5μm square, so the indentation test cannot possibly be measuring anything but the hardness of the fibre–matrix composite, whose plastic yield behaviour will be governed by the matrix. The included particles will have an effect on the yield behaviour of the matrix but are unlikely to govern it. Thus the measured hardness is must be that of the matrix. The matrix is presumably hardened in a manner similar to that of insect cuticle, but this is very much a matter for debate. The significance of the mineral inclusions now has a changed emphasis. The harder mineral is embedded in the softer matrix, suggesting that the matrix will wear away preferentially leaving the hard silica particles exposed giving a surface rather like that of sandpaper. A comment which doesn't necessarily upset the argument above is that the recorded hardness values for radular teeth are at least an order of magnitude too high! They are reportedly harder than the teeth of echinoderms, which are almost pure ceramic (Section 6.7). If the results are reduced to a tenth the reported value they come much closer to the values recorded for locust mandibles (Hillerton 1980), though still much higher by a factor of two. The reason may be either that the tests were done on sections of tooth which were too thin or stuck too firmly on the substrate, or simply that the instrument to measure hardness was wrongly used. Either way these experiments need to be repeated.

Runham's work attracted the attention of a number of people, notably RJP Williams of Oxford, who produced (with his students) a stunning series of studies showing the incorporation and distribution of silica and iron in a number of systems including the radula of *Patella* and other molluscs (Webb *et al.* 1989). Initial incorporation of iron is associated in position with brown colouration, presumably phenolic tanning since copper (found in phenolase) is also detected. Unfortunately none of this work correlated chemistry and ultrastructure with mechanical properties.

Iron and silica are not the only metals associated with hardening. Zinc and manganese have been found in a number of other animals, nearly always associated with the mandibles. Nearly all herbivorous insects have one or other of these metals in the cutting edge of the mandibles (Hillerton & Vincent 1982, Robertson *et al.*, 1984) where it is associated with increased hardness. It is present in relatively small amounts – about 5% dry weight of the whole mandible, so this represents much higher local concentrations which could possible be as high as those observed in the radula. The mandibles of the locust have a hard capping with a hardness of about $35\,\mathrm{kg\,mm^{-2}}$ (much lower than the radula, but it doesn't have to work against stone). The capping is on the inner edge of one mandible and the outer edge of the other. The soft faces (hardness about $18\,\mathrm{kg\,mm^{-2}}$) work against each other rather like a pair of scissors so that they wear each other away and continually break away the blunt edge by undermining (Hillerton 1980). In concept this is a bit like the hobby knife with a blade whose tip, when blunted, can be broken off to expose a fresh, sharp point. Some polychaete worms have zinc in the cutting edge and some spiders and scorpions can have Zn, Mn or Fe. The hardness has not been measured. The arachnids are unusual, compared with the insects, in that they are not herbivores and all three elements can be found in different bits of the mouthparts in a single individual (Schofield *et al.* 1989). Insects have either zinc or manganese, never (so far as we know) both.

5.6 CELLULAR MATERIALS

Plants, lots of bits of animals, bits of aircraft flying surfaces, helicopter rotor blades, skis, wasp nests and the doors on the underground trains in Hong Kong all have one thing in common – they are 'skin-core' or sandwich structures. A familiar one is the bracket fungus or polypore. The outer covering is a relatively stiff tensile membrane, but the inner material is very light and foam-like, made of cells. This form of structure has a number of structural advantages as well as the obvious biological ones. A few simple experiments bending the polypore fungus will convince you that the outer membrane is taking most of the stress. If you damage the skin the cellular core will fracture in tension relatively easily, although compressive failure is more controlled. It depends how dry the fungus is.

Gibson and Ashby have regularized much of the information about cellular materials (Gibson & Ashby 1988) starting from models rather like that shown in Fig. 5.24, which is an open-cell foam. When the foam is compressed the various components act as struts (i.e. loaded on end like a column) or beams. If the columns do not buckle out of the way (Euler buckling – see Gordon 1978) then most of the deformation will be due to the beams bending and the foam can be considered as a large collection of small beams. Standard beam theory (Timoshenko & Goodier 1970) gives the deflection of a beam of length *l* loaded at its midpoint by a force *F* as proportional to

$$F \, l^3/E_s \, I \qquad\qquad [\text{Eq.5.11}]$$

where E_s is the stiffness of the material used to make the beam and I is the

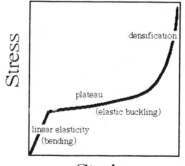

Figure 5.24 Stress–strain curve for compression of an open-cell foam made of elastomeric (rubbery) material (Gibson & Ashby 1988).

second moment of area of the beam, which in its turn is proportional to the fourth power of the thickness, t, of the beam, i.e.

$$I \propto t^4 \qquad \text{[Eq.5.12]}$$

The relative density of the foam is defined as the density of the foam divided by the density of the material making the beams – i.e. ρ^*/ρ_s. In turn,

$$\rho^*/\rho_s \propto (t/l)^2 \qquad \text{[Eq.5.13]}$$

(Gibson & Ashby 1988). The force is related to the compressive stress and is proportional to σl^2 and the strain for a given displacement is inversely proportional to l. Young's modulus of the foam becomes

$$E^* = (C_1 E_s \, I)/l^4 \qquad \text{[Eq.5.14]}$$

where C_1 is a rather convenient constant. Using Eqs 5.12 and 5.13

$$\frac{E^*}{E_s} = C_1 \left(\frac{\rho^*}{\rho_s} \right) \qquad \text{[Eq.5.15]}$$

So a log–log plot of relative stiffness against relative density has a slope of 2. This gives a nice way of checking the theory and seems to be true with a wide range of materials and foam shapes. However, this theory describes only the elastic part of the compressive behaviour – i.e. small strains. At larger strains all sorts of nasty things start to happen: the columns buckle and the beams bend plastically or break in a brittle manner. Eventually the bits get squashed so close together that they press directly on to each other again and the modulus starts to rise rapidly towards the stiffness of the solid material. This phase is known as densification. The precise behaviour in the zone between elastic behaviour and densification depends on whether the material from which the foam is made is rubbery, elastic-plastic or brittle. Figures 5.25 to 5.26 show the behaviour of three such materials in compression. In tension there is little qualitative difference between a cellular and a non-cellular solid in the overall shape of the curve, so an adaptation of Fig. 1.21 is adequate.

In the animal world the most studied cellular materials are ceramic (bones of one sort or another) which will be dealt with in the next chapter. But there are some keratinous cellular structures. The rachis of feathers is filled with a cellular foam which increases its stiffness by 16% (Purslow & Vincent 1978) as is the medulla of porcupine quills (Vincent & Owers 1986), although in the latter case the effect of removing the foam has not been investigated. The centre of the spines of hedgehogs (*Erinaceus* sp.) could also be considered as a sort of foam, although it is much more structured. It provides support for the wall so that the spine can buckle elastically without breaking (Vincent & Owers 1986). This may account for the numerous reports of hedgehogs being able to bounce, and the fact that they can have a semi-arboreal lifestyle without being able to fly or brachiate. Hedgehogs do not exist in the Americas, but their place may partly be taken by the tree porcupine, *Coendou*, which is unusual amongst the

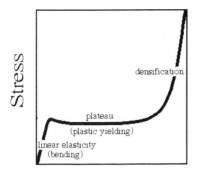

Figure 5.25 Stress–strain curve for compression of an open-cell foam made of elastic-plastic material (Gibson & Ashby 1988).

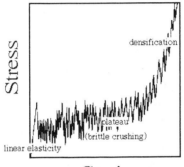

Figure 5.26 Stress–strain curve for compression of an open-cell foam made of a brittle material (Gibson & Ashby 1988).

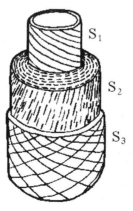

Figure 5.27 Primary, secondary and tertiary walls of a typical wood tracheid. Angles of the fibres in the various layers are S_1 (primary wall), 50–70°; S_2, 10–30°; S_3, 60–90°.

porcupines, but similar to hedgehogs, in having all its spines of the same length. Perhaps it, too, can bounce? Locusts make a cellular foam to plug their egg tubes, but these have not been investigated mechanically.

Cellular materials come into their own in the plant world, where the success of the design is very apparent. The most studied of these materials, mainly because of its commercial importance, is wood.

5.6.1 Wood

The specific stiffness of wood (stiffness per unit weight) is as good as that of steel. Moreover steel costs about sixty times more than wood per ton. Wood is a quarter as strong as mild steel so its specific strength is about four times that of steel. So its commercial importance is great – if only we could stop it from rotting. But unfortunately the techniques used to stop the rot very often reduce the strength of wood. Also because of its complexity, wood is not always an easy material to deal with. It needs a much more skilled and sympathetic approach than do metals (Gordon 1976).

Many studies on the structure of wood show it to be composed of parallel columns of cells. These cells have the cellulose wound around them spirally, mainly in one direction (Fig. 5.28) embedded in a matrix of lignin, a complex polymeric resin which has some resemblance to the polyphenols found in insect cuticle. In tension, and to a first approximation, the cells can be considered as a system of parallel discontinuous fibres embedded in a matrix. This does not take into account the fine morphology of the cells but in practice this seems not to matter. For instance, for sitka spruce (*Picea sitchensis*) the mean axial tensile stiffness and strength of single cells are 12.5 GPa and 170 MPa, respectively (referred to the gross cross-sectional area). These figures can be put into formulae like those for fibrous composites (Eqs 5.1 and 5.2), neglecting the contribution of the matrix:

$$E_c = E_f V_f \qquad [\text{Eq.5.16}]$$

$$\sigma_c = \sigma_f V_f \qquad [\text{Eq.5.17}]$$

The volume fraction of cells in sitka spruce is about 80%, so the calculated values for E_c and σ_c are 10 GPa and 136 MPa which agree well with the mean experimental values of 10 GPa and 100 MPa. To some extent this might imply that the cells are to be considered as separate entities from the mechanical point of view. In an exhaustive study of the mechanics of wood, Mark (1967) showed that this is so and that a single wood cell has different behaviour from the same cell cut in half longitudinally. This emphasizes the importance of the cellulose and its continuity for the mechanics of the cell. Alternatively one could consider the cellulose microfibril as the basic unit of the composite and derive the

Figure 5.28 Wood tracheids showing how the spirally wound cellulose in the S2 wall (Fig. 5.27) splits and allows the tube to buckle inwards. The tracheids are about 20 μm in diameter (courtesy of Dr Giorgio Jeronimidis).

mechanical properties from the information available on the laminated structure of the cell wall. This is a complex problem, but Mark has given a comprehensive account of this approach, as has Jeronimidis (1980).

The major wall layer in the wood cell wall is the S2 layer which contributes up to 80% of the total thickness. It is therefore also the major load-bearing component. Gordon & Jeronimidis (1980) have analysed the way this layer contributes to the toughness of wood and, in doing so, invented a new type

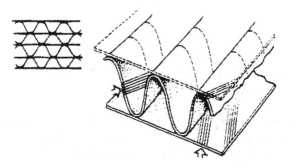

Figure 5.29 A further refinement of the wood model leads to this material rather like corrugated cardboard. The fibre directions give the entire material some of the characteristics of wood (Chaplin *et al.* 1983).

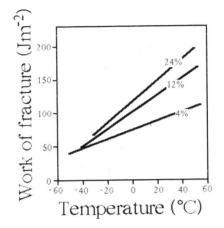

Figure 5.30 Variation of toughness of wood with temperature and water content, cracking *along* the grain. Figures against the lines indicate the water content (g g dry weight⁻¹) (Jeronimidis 1976).

of glass-fibre–resin composite (Fig. 5.29). The effect of temperature on the toughness of wood can provide some clues: at a constant moisture content the energy required to propagate a crack along the grain increases with temperature (Fig. 5.30). This implies that the fracture energy of wood for cracks parallel to the grain is due to the matrix component behaving like a semi-brittle polymer. Across the grain, the relationship is different (Fig. 5.31) and a decrease in temperature increases the work of fracture. This result has been repeated by a number of people, which gave Jeronimidis more confidence in this result, since it is rather strange: most materials show a decrease in toughness with decreasing

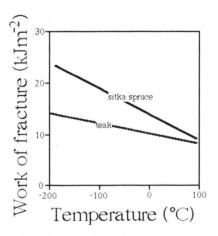

Figure 5.31 Variation of toughness of wood with temperature, cracking *across* the grain. Water content is 12% g g dry weight⁻¹ (Jeronimidis 1976).

temperature. This is probably due to improved transference of stress to the crack tip – less energy will be lost in viscous processes at lower temperatures, since the molecular relaxations leading to those viscous processes will be less liable to occur. The reasons for the decrease in toughness across the grain as temperature is lowered is not clear but this behaviour is important in relation to the physical mechanisms for energy absorption during fracture. Also, since a great number of the world's forests grow at latitudes where temperatures of $-30°C$ are fairly common, a toughness mechanism which is little altered by temperature is clearly an advantage for the trees.

   If one calculates the toughness of a fibrous composite using the same volume fractions of fibre and matrix as wood, with the fibres arranged as in a normal glass-fibre–resin composite and assuming that the major toughening mechanism is the friction between fibre and matrix as the fibres are pulled out, the figure for work of fracture is ten times less than is actually observed with wood. Moreover one does not see many fibre ends in the broken surface, which would be expected from a 'normal' glass-fibre – resin composite. For a system of parallel discontinuous fibres, much like that of the locust apodeme, the work of fracture due to pull-out of fibres can be expressed as:

$$W_{(\text{pull out})} = \frac{1}{48} \frac{V_f \sigma f^2 d}{\tau} \qquad [\text{Eq.5.18}]$$

where the symbols are the same as above; in addition d is the diameter of the fibres, τ, is the interfacial shear strength between fibre and matrix. Table 5.5 shows a comparison between fracture energy values calculated with Eq. 5.17 and those measured experimentally for softwoods and hardwoods.

   It appears that the mechanism of fracture leading to this apparent anomaly is due to the helical arrangement of the fibres in the load-bearing S2 wall. If a single wood tracheid is tested in tension with the ends prevented from rotating relative to each other, the fibres shear apart and buckle inwards into the lumen of the cell. This not only gives a higher work of fracture but also gives a stress–strain curve more like that of a metal (Fig. 5.32) with a distinct yield point and a post yield region where the material, although having failed, is still capable of supporting a load. The features of this type of failure, which have been observed, are that the cracks in the S2 wall run parallel with the fibres, that the S2 wall decreases locally in diameter at the early stages of buckling and

Table 5.5 Fracture energy of hardwood and softwood

	σ_f (MPa)	V_f (%)	d $(10^{-5}\,\text{m})$	τ (MPa)	kJm^{-2} calculated	measured
Soft	180	0.96	3.0	12	1.6	12
Hard	240	0.73	1.5	14	0.94	11

From Jeronimidis (personal communication)

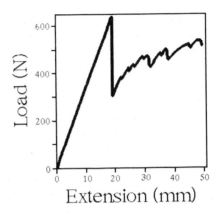

Figure 5.32 Load–extension curve for a single 'straw' of wound glass fibre in epoxy resin matrix. The fibres are wound at an angle of 15° to the longitudinal axis; the maximum corresponds to the onset of buckling (Gordon & Jeronimidis 1980).

separates locally from the outer primary wall (the creation of this new surface absorbs more energy) and that later stages of failure cause the S2 wall to fold inwards. It seems that the orientation of fibres in the S2 wall is something of a compromise between stiffness and toughness. If the fibres are orientated more parallel to the cell axis (as happens in grasses) the cell is stiffer but cannot take advantage of the buckling mode of failure and so is less tough. This is not so important for an annual plant, which anyway is not exposed to the same sort of external loads as a tree. But if the fibres are orientated more at right angles to the axis of the cell the toughness will be increased but the axial stiffness reduced. If the compromise is well calculated it is possible, in artificial composites using the same morphology, though on a much larger scale,

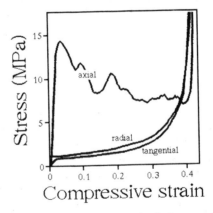

Figure 5.33 Compressive stress–strain curves for balsa wood. Orientation is relative to the long axis of the stem and the wood cells (Easterling *et al.* 1982).

to increase the toughness by a factor of ten or more over 'traditional' glass-fibre–resin composites at the expense of only a moderate decrease in axial stiffness. Early forms of these models were assemblages of straws of spirally wound glass fibres stuck together with resin (Gordon & Jeronimidis 1980). But these are not very easy to wind and stick together. In order to make a cheaper analogue, a paper-making company was approached. Paper is a fibrous composite which has been made for hundreds of years, so there is plenty of experience available in manipulating such things as the relative orientation of fibres and the amount by which they are stuck together. It turns out that with a bit of imagination you can make a wood analogue which looks very much like corrugated cardboard (Fig. 5.29) but which, because of the orientation of the fibres, fractures rather like wood and has greatly increased toughness and resistance to impact. The general idea of using biological materials as models for man-made materials is not a new one and will be returned to in the next chapter.

The compressive failure of wood is best modelled by treating it as a cellular material. Once again its mechanical behaviour is very anisotropic. When squashed sideways the cells collapse pretty quickly (Fig. 5.33), but end-on they can take quite a respectable load though not so much as in tension. So a bent wooden beam will fail first in compression. Ultimately the S2 wall creases at an angle to the grain direction. Such a failure is perfectly safe so long as the wood is not put into tension. If this latter occurs then the crease will function as a crack and concentrate the stress leading to premature failure. That is why it is not a good thing to land a glider heavily or stop a rowing boat with the back of an oar: in both instances the normal stresses are suddenly reversed and the compression face is subjected to tension – with possible catastrophic results. The story is told of a swimming pool attendant who thought that the springboard was looking rather old and worn. The under surface was still quite pristine, so one night he turned the board over. Next morning the first person to use the board found that quite a few compression creases had accumulated, safely, on what had been the compression side of the board – only it was now the tension side.

5.6.2 Other plant structures

Although the general idea that non-woody plants achieve their compressive stiffness through turgor stiffened cells is well known, no satisfactory model has yet been produced despite much effort. Probably the best in terms of general applicability is by Gibson *et al.* (1988) for the iris leaf. This adapts the general ideas of cellular materials and sandwich structures to produce a model which predicts the stiffness of the leaf to within a factor of 2 or less. The nice thing about this model is that it is relatively simple, although it achieves this simplicity by side-stepping such problems as admitting turgor pressure as a variable. There are many models which have attempted to account for the variation in stiffness of cells as a function of turgor (e.g. Nilsson *et al.* 1958; Wu *et al.* 1988), but

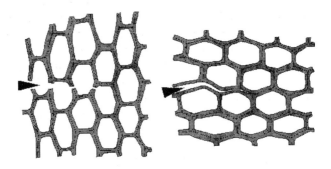

Figure 5.34 Different ways for cellular material to break in tension: (left) when the cells are well stuck together and the cell walls relatively weak; (right) strong cells adhering weakly (Gibson & Ashby 1988).

I have yet to see a successful one which starts from the mechanics of a single pressurized cell and integrates it into a complete plant organ such as a leaf or stem. This shortcoming has been pointed out by Steudle & Wieneke (1985). Part of the problem which has not been properly addressed is that the cells are not stuck together totally rigidly. The degree of cell–cell adhesion can change both with turgor and with maturity. As turgor drops, so the cells are pressed together less and are less capable of transmitting shear forces, because both 'friction' and the area of mutual contact are reduced. For instance the stiffness of apple tissue can drop by a factor of 2 or 3 simply due to reduction in cell–cell adhesion (Vincent 1989). And as the degree of cell–cell adhesion changes, so does the mode of fracture (Fig. 5.34). The cells of the potato tuber are well stuck together with little space between, so the cell walls tend to fracture preferentially. Apples, by contrast, have large air spaces between the cells (at least, radially) and tend to fracture between the cells. This is especially noticeable in over-ripe or 'mealy' apples: the cells are about 0.1 mm in diameter and easily separated, so they can be rubbed off with the fingers when they feel rather like sand grains. The fracture of plant materials is a fascinating area which has been mostly very poorly researched, despite its commercial importance (Vincent 1990). Animals need to fracture plants at cellular, tissue and organ levels in order to feed on them; many plants shed leaves and twigs as part of their growth and development; seed pods dehisce; nuts give protection. The quality of research in this area may be expected to improve partly as people become more aware of its existence and partly as the theory of cellular materials is developed further to a better understanding of fracture and the inclusion of turgor effects.

It is possible to model cellular systems using a parameter analogous to the density ratio of Gibson & Ashby by considering the cellular material to be uniformly distributed rather than concentrated in cell walls. Thus when, as in the stems of many herbaceous plants, the size and wall thickness of the cells varies across the stem, this is represented by a gradient in volume fraction of cell wall material. The turgor effects can then be modelled as a generalized pre-stressing of the cell wall material. For a given area of stem, the prestress *in the stem* will be the product of prestress and volume fraction. Thus stem prestress will be greater around the outside of the stem where the cells tend to be smaller and with thicker walls (Vincent & Jeronimidis 1990). On the whole, however, the analysis of cellular materials has not at the time of writing (1990) progressed to the stage of modelling herbaceous plants. And what happens in plant stems which contain sclerenchyma fibres running along their length?

5.7 MATERIAL OR STRUCTURE?

One of the big problems when dealing with biology from an engineering point of view is to know whether one is dealing with a material or a structure. Hair, horn, cuticle and wood all have "structure" and yet are mostly treated as materials. In fact it is not until the differentiation between material and structure can be made that such complex materials can be at all understood. It is still easy to find papers newly published which talk about the stiffness of a leaf as if the leaf were uniform in structure. There is an easy test. If you are dealing with a material then it will have the same stiffness in tension, bending and compression. If it doesn't, then you must be dealing with a structure and the stiffness which you are measuring is not a material parameter. You will then have to look more closely and attempt to model the structure, as has been illustrated in this chapter. Of course, there is still no guarantee that you are *not* looking at a structure if stiffness appears to be constant, but it is a pretty good test. But how much structural top-hamper do you have to put up with when testing a biological 'material'? Is there a rule of thumb which says 'below a certain size level of the components you don't have to think about the structure of the material but can consider it as homogeneous, described by a few numbers'? I don't know that there is, but it seems that it is often possible to identify a size threshold below which the structure of the material does not affect the mechanical properties. This threshold moves depending on the size of the testpiece, the conditions under which the test is performed and the material in question. It is determined, it seems, by observing the response of the testpiece and considering at what level in the size hierarchy the main interfacial influences occur. For instance in studying fracture of bone (Chapter 6), cow leg bone is so well mineralized that the main functional unit is the osteon, whereas antler bone is less well mineralized so that the main functional unit is the collagen-hydroxyapatite fibre.

Ultimately you can always fit a polynomial function to whatever properties you are measuring, or apply some other 'black box' mathematical function. Some well-respected researchers are still doing this sort of thing, but unless you simply want something equivalent to a 'quality control' test and are not concerned with understanding the material you are investigating, this 'constitutive equation' approach does not seem to be a very productive one.

CHAPTER 6

Biological Ceramics

The trouble with using protein to make a skeleton is that it is expensive. Insects use it because it is relatively light and, with chitin as the fibre, can make a stiff and tough skeleton. Moreover its mechanical properties can be very closely tailored to its use by varying the properties of the protein matrix, so that insect cuticle can be adapted to perform many functions, from making hard mandibles to elastic and extensible membranes. Protein is thus eminently suitable for the skeleton of an animal which owes a large part of its success to its capacity for flight. But protein has to be synthesized – an energy-consuming process – and is largely placed outside the metabolic pool by the necessity for mechanical stability. The more it is cross-linked the less soluble it becomes. Crustacea – arthropods which are mainly aquatic and do not fly – reduce the amount of protein in the exoskeleton and replace it with calcium carbonate. The specific gravity of calcium salts is more than twice that of protein and chitin, but since this in part reflects closer packing of the molecules (and hence greater bond density and greater stiffness, E, of 100 GPa) it is not patently obvious why the calcium-containing exoskeleton should be heavier. A likely explanation is that it is much more brittle and that under a given force a thicker skeleton will be far less likely to approach its ultimate strain because the stress will be that much less. The insect can control the deformability of its skeletal material much more closely – but the (mainly) calcite of decapods is brittle because it is crystalline and this property cannot be changed, so its bulk fracture properties can be circumvented only by careful design of the material. How can animals and plants circumvent the low fracture toughness (and hence general durability) of ceramic materials in order to make proper use of their greater stiffness and strength? One particularly elegant example is found in the calcified cuticle of one of the forelimbs of a stomatopod crustacean, *Gonodactylis chiragra* (Currey *et al.* 1982). The limb is used to break mollusc shells in impact, being powered by a catapult system. The outer layers reach an astounding 1000 kg mm^{-2} hardness which is way over most of the values quoted in Table 6.1. It does this by including phosphate in the normal calcite of the exoskeleton, though (as Currey points out) this might be reinforced in its effects by the fact that the samples were kept in alcohol for a number of years, which would tend to remove

Table 6.1 A comparison of calcium-containing minerals, some of which are used in animal skeletons and other hard parts

Chemical formula (a)	Common name (a)	Occurrence (b)	Density (a)	Hardness (Mohs) (a)
$CaCO_3$	Calcite	Birds' eggs Echinoderm test, teeth Octocorallian spicules, Sponge spicules, Some brachiopods, Crustaceans, Molluscs	2.71	3
	Aragonite	Some reptile eggs, Some Foraminifera, Many molluscs	2.93	3.5–4
$CaMg(CO_3)_2$	Dolomite	Echinoderm teeth	2.85	3.5–4
$MgCO_3$	Magnesite	Sponge spicules	3.01	4
$Ca_5(PO_4)_3(OH)$	Hydroxyapatite	Bone, Teeth, Dermal ossicles, Young molluscs	3.1–3.2	5
$SiO_2.(H_2O)_n$	'amorphous' hydrated silica (Opal?)	Sponge spicules, Limpet radular teeth	2.0–2.2	5.5–6.5
CaF_2	Fluorite		3.18	4
$CaSiO_3$	Wollastonite		2.9	5
$(Ca,Na)(Al,Si)AlSi_2O_8$	Plagioclase, Scapolite		2.6–2.8	6
$Ca_2Al(AlSi_3O_{10})(OH)_2$	Prehnite		2.9	6.5
$CaAl_2Si_2O_7(OH_2).H_2O$	Lawsonite		3.1	8

From (a) Mason and Berry (1968); (b) Brown (1975)

water and hence remove some plastic deformation. Even so, it is an impressive performance.

In a biological ceramic, the control of the manner of growth of the crystals is obviously important. Even more so when one realizes that the controlled crystallization of inorganic materials in the laboratory is a black art which is neither predictable nor explicable, yet it is one of the oldest of the chemist's (or alchemist's) arts. So whilst biologists (as usual) take it for granted that animals and plants can manipulate the mineral phase in ceramic composites and ask questions about the physiology and mechanics, the chemists are still posing the question 'How?' To a large extent this particular question is beyond the scope of this book, fascinating as it is. Refer to Lowenstam & Weiner (1989) and to Mann *et al.* (1989). When all the chemistry has been chewed over, the frustration of the chemists can be judged by the readily observable fact that the mineral phase in biological ceramics is controlled in terms of the size, shape, orientation, chemical composition, crystal type and degree of bonding within the matrix. With control like this it is perhaps not surprising that biological ceramics can perform at least as well as the best man can make, and usually much better. Their only weakness is the inability to stand high temperature because the protein denatures at 60°C or so.

6.1 CALCIUM SALTS OR SILICA?

One advantage of using calcium salts seems to be metabolic cheapness. Unfortunately the metabolic cost to an animal of making something out of calcium salts has not been calculated. Certainly calcium can be relatively easy to come by. Sea water, especially along the tropic strand, is saturated with calcium. Two major classes of calcium salt are used in skeletons – phosphates and carbonates. These different salts have different crystalline forms and this introduces an additional factor in the variety of materials possible. In general, phosphates (as hydroxyapatite) occur in conjunction with collagen, carbonates (as calcite, aragonite, etc.) in conjunction with other proteins and polysaccharides. Phosphates also tend to be the major form in vertebrates and brachiopods, carbonates in most other invertebrates. There are other salts such as oxalates but these are not common. Obviously such salts are largely insoluble once formed and precipitated. Some of these salts are listed in Table 6.1 together with an indication of their distribution in the animal kingdom, their density and their hardness on the Mohs scale (Fig. 6.1).

The importance of hardness is not so much in that property itself (though it probably is important in abraded materials such as tooth enamel) but because the tensile strength of a crystal is a maximum of one third its Vickers hardness number (less if dislocations are free to move within the crystal and increase the amount of plastic deformation). Thus the hardness of a crystal can give an indication of the strength of the material it comprises, which is very useful when the material is in a shape or size which is difficult for direct tensile tests. The correlation is possible because hardness is a direct function of the energy of the interatomic bonds and the type of bonding. This has been shown for a variety of minerals: in general minerals whose atoms are bonded covalently, such as silica, aluminium oxides and diamond, are very much harder than minerals whose atoms are bonded ionically, such as sodium chloride and calcium carbonate. Hardness is also affected by the symmetry of bonding – the plates of graphite are actually harder than diamond but they can slip over each other and so form a soft material.

Another factor affecting hardness is the size of the ions or atoms involved. The smaller the ion the denser the packing and the denser the bonds in the crystal. Hence higher stresses can be borne. For the same charge the ions of manganese, iron, cobalt, nickel, magnesium and beryllium are smaller, and in that order, than calcium. So a calcium salt with any of these ions incorporated into it will

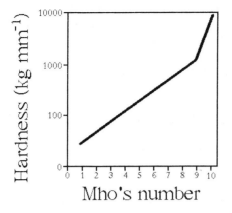

Figure 6.1 Relation between Mohs number and indentation hardness. The numbers on the Mohs scale refer to the following standard materials: 1, talc; 2, gypsum; 3, calcite; 4, fluorite; 5, apatite; 6, orthoclase; 7, quartz; 8, topaz; 9, corundum; 10, diamond (Mott 1964).

be harder. Echinoderms make use of this by incorporating dolomite into their teeth (Brear & Currey 1976) and some fish incorporate iron into the hydroxyapatite of their teeth (Motta 1987). Another way of expressing bond density is the density of the material (assuming the chemistry does not change); Table 6.1 shows that aragonite gains its greater hardness by closer packing of the ions and that apatite is probably also more tightly packed within the crystal structure. This comparison of density and hardness also shows that the silicates, which are largely covalently bonded, are less dense than the ionic salts at equivalent hardness. So if hardness (or, more probably, strength) is at a premium, then silicates will be the choice. Why should silicates be so rare? Probably the answer is to do with the chemistry of silicon. It may be metabolically more expensive to incorporate silicates; certainly silicates are in general less soluble than calcium salts. The cycle of silica in nature is thus complicated by the almost complete removal of assimilated silica from the pool of available nutrients when the organism dies or is eaten. By contrast some calcium salts (e.g. sulphate, chloride) are much more soluble than others (e.g. carbonate) so that Ca^{2+} is not only readily available but can be made insoluble when sequestered.

But when all is said and done there is not necessarily any advantage in silica for skeleton or spicules. A consideration of the strength of whiskers of zinc oxide and silica (Fig. 6.2) shows that size is as important as anything and that the advantages to be gained in extra strength can easily be lost, as with glass fibres, by having crystals which have growth steps in them which can act as stress concentrators. These steps will then initiate cracks and cause the crystals to fail in a brittle fashion well below the theoretical strength (Gordon, 1976). This problem seems to have been circumvented by a remarkable structure

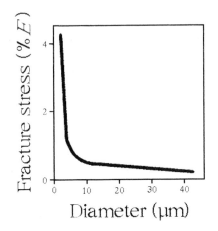

Figure 6.2 Relationship between strength and diameter for non-cleavable whisker crystals such as silicon or zinc oxide (Cook & Gordon 1964).

produced by a marine sponge. *Monoraphis* is a rare sponge with a single anchoring spicule several millimetres in diameter and up to a metre long (Levi *et al*. 1989). The spicule has a modulus of about 36 GPa (about half that of a vitreous silica rod) and seems to be made of some sort of opal, the form in which silica seems to be found throughout the animal and plant world. The spicule can be bent into a circle without breaking. It seems to be able to do this because it is made up of a large number of concentric layers – over 200 in the specimen examined – varying from 10 μm thick in the middle to 3 μm at the periphery. Comparing load deflection curves of a spicule and a silica rod of similar size in three-point bending, the spicule can achieve seven-fold higher deflection because the layered structure can control the propagation of cracks. It is thus able to achieve a four-fold higher stress before failure (593 MPa as opposed to 155 MPa), and even then breaks in a relatively controlled fashion. This gives it a 30-fold greater work to fracture.

6.2 MOLLUSC SHELLS

Amongst the invertebrates calcified tissues are found in most, if not all, the phyla in one form or another. Of these the most easily, and therefore the most extensively, investigated are the shells of molluscs. There are six main types of shell material (Table 6.2) which show differences in structure and composition; nacreous (Fig. 6.3), prismatic (Fig. 6.4) and crossed lamellar (Figs 6.5, 6.13) are illustrated. These types of material can occur alongside each other in a single shell; in bivalves, for instance, there are three characteristic arrangements: (a) nacre inside, prismatic material outside; (b) foliated material with a little crossed lamellar material; (c) crossed lamellar material, either on its own or with a little

Table 6.2 Types of mollusc shell material

Type	Crystals		Protein matrix (% by weight)	Shape	Mean strength (MPa) (Max. values below mean unless only one determination available)			Stiffness (GPa) (no. of determinations in parentheses adjacent, max. below)	Vicker's hardness number (kg mm^{-2}) (with no. of determinations)
	Calcite	Aragonite			tension	compression	bending		
Prismatic (Fig. 6.4)	*		thick (5 μm) sheet around each prism (1–4%)	polygonal columns, 100–200 μm × several mm long	60 (60)	250 (300)	140	30 (2) (40)	162 (1)
Nacreous (Figs 6.3, 6.8, 6.9)		*	thin layer between sheets (1–4%)	flat tablets in sheets (0.3–0.5 μm thick)	130 (wet) 167 (dry)	380 (420)	220 (280)	60 (wet) 70 (dry)	168 (8)
Crossed lamellar (Figs 6.5, 6.13)		*	very tenuous (0.01–4%)	plywood-like lamellae (20–40 μm thick)	40 (60)	250 (340)	100 (170)	60 (14) (80)	250 (9)
Foliated	*		very tenuous (0.1–0.3%)	long thin crystals in overlapping layers	30 (40)	150 (200)	100 (180)	40 (6) (60)	110 (2)
Homogeneous		*	very tenuous	fine-scale rubble (0.5–3.0 μm diameter)	30	250	80	60 (1)	—

From various papers by J. D. Currey

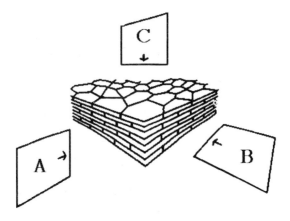

Figure 6.3 Orientations for fracture in nacre: A and C are 'difficult' directions since they cut across plates. B is an 'easy' direction between the plates.

homogeneous material. Other combinations of material type are found within a single shell but these three are much the commonest. The type and arrangement of material tends to be consistent with phylogeny, usually all species within a family having the same arrangement, sometimes even those within a superfamily. Thin shells tend to be made of prismatic material with nacre or foliated material; very thick shells tend to be made of crossed lamellar material. Surprisingly enough, nacre is probably the most primitive of the shell materials, and is composed of an intrinsically weaker ceramic. Currey has studied the strength of nacre and crossed lamellar shell and has shown that the governing factor is the structure of the material (Currey 1977).

When it comes to measuring the mechanical properties of these materials it is very important that the test conditions are correct. This may seem an obvious thing to say but it is by no means an obvious thing to do! Many tests to measure stiffness and strength are performed on a beam in three-point bending. This is especially convenient since loads can be reasonably easily fed in to the specimen, compared with the complexities of end fixings for compressive and tensile tests. However, if the beam is short relative to its depth, it will experience a significant amount of shear through its thickness. This can be accounted for if you know enough about the material, but it is far more reliable to avoid this complication. Suitable span-to-depth (S/D) ratios, chosen to avoid shear, have been recommended for various materials including metals (S/D > 8), timber (S/D > 24) (Roark & Young 1975) and fibre-reinforced composites (Zweben *et al.* 1979) but none has been given for nacre. The importance of getting this ratio right is shown by experiments with a unidirectional Kevlar-polyester composite (Zweben *et al.* 1979): with the use of S/D = 16 instead of the required 60, the measured modulus drops to about 65% of its true value! With nacre, a S/D ratio of at least 16 is found to be necessary (Jackson *et al.* 1988) when the values shown in Table 6.3 are obtained. These values may be

Figure 6.4 Fracture surface of prismatic shell material (courtesy of Kevin Brear and Professor John Currey). Scale = 12 mm = 50 μm.

higher than the values obtained by Currey simply because he used too small a S/D ratio.

In order to understand the way nacre works as a material we need to be able to model it, which will involve composite theory again. Taking the stiffness of

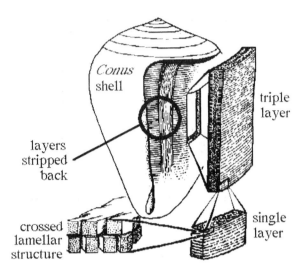

Figure 6.5 Shell of *Conus* showing the structure of crossed-lamellar shell.

aragonite as 100 GPa (Hearmon 1946) and that of the closest material to the matrix, keratin matrix from fresh horn, as 3 GPa (Kitchener & Vincent 1987) the Voigt model gives a Young's modulus of 95 GPa and the Reuss model, 39 GPa. For the dry material, taking the matrix modulus as that of dry keratin matrix (i.e. 6.1 GPa), the Voigt estimate is hardly affected; the Reuss model gives 56.5 GPa. Thus neither model predicts the actual modulus of nacre very well. However, these two models make the assumptions that the Poisson's ratio of both components is identical and that the fibre (= ceramic) phase is composed of infinitely long particles. Padawer & Beecher (1970) have produced a model which gives an equation rather like Eq. 5.1 but in which the value of z is different. There is a similar model by Riley (1968) modified by Lusis *et al.* (1973). An additional parameter is the shear modulus of the matrix material, which can be measured on a short beam (S/D = 3 or so). It comes out as 4.6 GPa (dry nacre) and 1.4 GPa (wet nacre) (Jackson 1986). When the Padawer-Beecher and Riley models are calculated over a range of matric shear moduli, the curves

Table 6.3 Some mechanical properties of nacre

Hydration	S/D	Toughness (J m^{-2})	Ductility (%)	Stiffness (GPa)
desiccated	4	264	0	
dry	16	352	19	73 ± 9
	4	464	29	
wet	16	587	44	64 ± 8
	4	1240	55	

From Jackson *et al.* (1988)

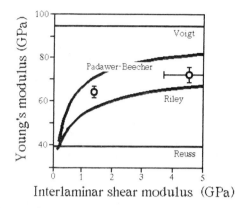

Figure 6.6 Estimates for the Young's modulus of nacre given by Reuss, Voigt, Riley & Padawer-Beecher models (see text). The circles are experimental data points with standard deviations (Jackson *et al.* 1988).

shown in Fig. 6.6 are obtained (Voigt and Reuss values are shown for comparison). The measured values for wet and dry nacre are also shown (Jackson *et al.* 1988). Composites made with platelets of mica fall well below the Riley prediction. From this and other arguments (involving the assumptions used in the Riley and Padawer-Beecher models) nacre can be taken as a near-ideal platelet composite (Jackson *et al.* 1988). Further modelling suggests that the strength of nacre can be best explained by the plates pulling out from between each other, rather than breaking (Jackson *et al.* 1988). This is confirmed by models made with glass microscope slides stuck together with epoxy resin (Jackson *et al.* 1989).

Nacre loaded in tension is initially Hookean, then plastic up to strains greater than 0.01 (and locally probably higher) (Fig. 6.7). In the plastic region the material goes milky in appearance producing an appearance very much like the craze that appears in polymers strained near their breaking load (Currey 1977). This observation is returned to later. When the material breaks, its behaviour is very different depending on whether the crack travels across (the strong & tough direction) or between (the weak direction) the platelets. The work of fracture varies between 200 and 1250 Jm^{-2} depending on S/D ratio, degree of hydration and orientation of the crack path (Table 6.3). Several features combine to give this toughness.

1. The sheets are about 0.5 μm thick, but the critical crack length is about three times this so that the sheets are not thick enough to contain dangerous flaws which could start a crack running through the material (Currey 1977).
2. Ninety-five per cent of the sheets are within 30% of 0.5 μm thick so that there are no odd sheets which could contain a crack of critical length (Currey 1977).
3. The fracture is continually deflected to each side of its main path (Currey 1977). In addition many plates are separated laterally (Fig. 6.8) (Jackson *et al.* 1988).

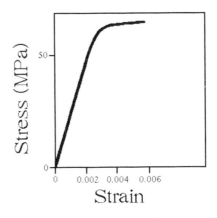

Figure 6.7 Tensile stress–strain curve for nacre (Currey 1980).

4. A fracture which does not go through a sheet of ceramic goes through the matrix, indicating that matrix and ceramic phases are well bonded (Currey 1977).
5. When the matrix is fractured, it forms many threads which connect the two surfaces (Fig. 6.9) (Jackson *et al.* 1988).

Clearly the matrix contains some secrets. It also contains soluble proteins rich in acidic amino acids (aspartate and glutamate present in their acid form) and acidic sulphated polysaccharides; insoluble hydrophobic proteins rich in glycine and alanine and also β-chitin. So the matrix is then envisaged as a super-sandwich structure with chitin in the middle (Fig. 6.10). The acidic groups are considered to direct the nucleation of the aragonite sheets (Weiner & Traub 1984). The presence of co-ordination sites at the interface involving Ca, N, O and S was shown using advanced techniques for analysing the chemistry of surfaces (Jackson *et al.* 1987). So the matrix is well bonded to the ceramic phase. When the matrix is cleaved, the threads which are formed are very similar to the fibrils which occur in crazes (thin, planar, defects which strongly reflect and scatter light) which form during ductile failure in thermoplastic materials (Kramer 1986). In the fibrils, the polymer molecules are in all probability orientated in the direction of extension effectively connecting the two surfaces between which they are stretched and enabling the craze to support tensile stresses (Kramer 1986). So it seems very likely that the threads of matrix in nacre are spun from the chitin and silk-like proteins postulated by Weiner & Traub to give an exact analogue of craze fibrils, allowing even fractured nacre to continue to carry loads. Additionally, the formation of new surface within the matrix as the crack goes through it, as well as the formation of the threads, will absorb large amounts of fracture energy, contributing to the ductility of nacre. Under suitable conditions this ductility can account for more than half the work of fracture of nacre (Table 6.3).

When nacre is built into a shell it is laid down so that the sheets are parallel to the surface of the shell. Thus any crack travelling through the shell traverses

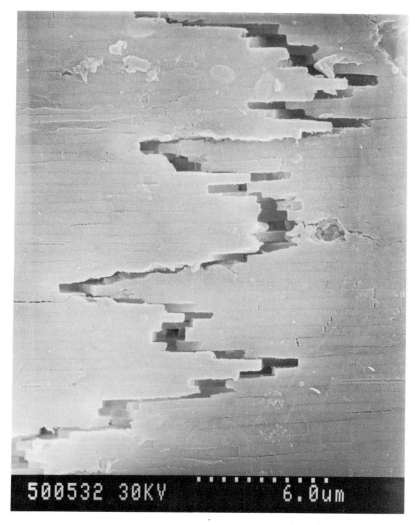

Figure 6.8 Fracture path through nacre viewed in orientation A (Fig. 6.3) (Jackson *et al.* 1988).

the nacre across the platelets in the direction of greatest toughness. In the intact mollusc, catastrophic failure is going to be very likely in whole, undamaged shells which will behave like un-notched beams at a large S/D ratio. Furthermore, control of crack propagation is made more difficult since the muscles of a predator will store significant amounts of strain energy and operate like a 'soft' machine. You experience this every time you crack a nut between your teeth. Without sophisticated control mechanisms in your jaw muscles your teeth would come together very hard when the nut shell finally breaks. Thus by the time a predator has reached the stress required to break a shell, it will

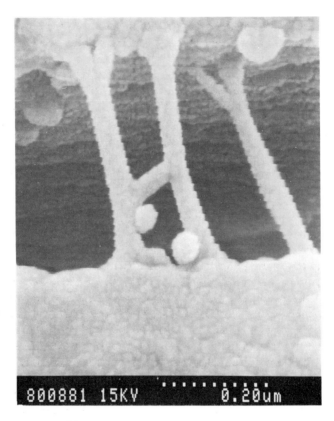

Figure 6.9 Close-up of fibres extending across a craze-like fracture between plates (Jackson *et al.* 1988).

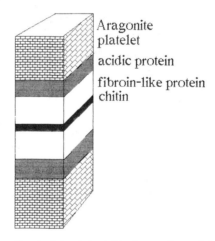

Figure 6.10 Model for the matrix material between two plates of nacre, showing acidic and fibroin-like proteins with chitin centrally (Weiner & Traub 1984).

have large amounts of strain energy ready to feed into the advancing crack. When failure does occur it will be lethal to the mollusc because outer shell materials, by all accounts, are not sufficiently strong or tough enough to arrest a fast-moving crack. Currey observed crack stopping in nacre, but only by reversing the normal direction of bending (Currey 1977). Because whole shells will normally break catastrophically, what is most important to the organism is not the detail of crack propagation, as Currey implied, but the amount of additional energy absorbed through ductility processes before fracture actually starts. The absorption of energy during the propagation of a crack may not be an option open to the organism, because once the failure load has been reached, drastic measures are needed to arrest a crack that is being fed by a large amount of strain energy from the predator's muscles. This may be the reason why some species of mollusc have sacrificed the high work of fracture of nacre and have opted instead for a strategy whereby cracks can be stopped before it is too late to do repairs. If pieces of broken shell become too widely separated (as they would after catastrophic failure) then it is impossible for the animal to lay down material from the inside to heal the gap. Molluscs containing crossed-lamellar material (see below), however, have the advantage of a criss-cross arrangement of laths within their shell which enables cracks to be stopped very effectively (Currey & Kohn 1976). The effect of the S/D ratio will also be of interest to molluscs with long thin shells as opposed to those with short fat ones. If, for example, shells grow more in length than in thickness then young shells with a small S/D ratio should be more ductile and able to absorb more energy per unit mass of shell than can old ones. Perhaps this is the reason why so many molluscs have shells made only of nacre when they are young, but add other types of shell material when they get older and larger.

The other type of shell material whose fracture behaviour has been studied is crossed lamellar. This is commonly found in the shells of *Conus* species (Fig. 6.5), where there are usually three layers of differing orientation. The lamellae in the middle layer run parallel to the lip and through the thickness of the shell, so lie in the *zy* plane. The lamellae in the outer and inner layers run towards the lip and through the thickness and so lie in the *xy* plane. The specimens providing the crossed lamellar material quoted in Table 6.2 had their longitudinal axis in the *z* direction and this should be noted when comparing the following with the results in that table.

The load-deformation curve of *Conus* shell in bending (Fig. 6.11; Currey & Kohn 1976) is saw-toothed initially, followed by a smooth region before failure. The saw-teeth have been found to be due to cracks running through the tensile layer between the lamellae and then being stopped by the middle layer; the following smooth region correlated with the slight extension of one of these cracks into the middle layer; failure due to the sudden propagation of this crack through the entire middle layer (Fig. 6.12). There are two major explanations for the difference in fracture behaviour between the two layers: either the middle layer is chemically and/or morphologically different or its properties arise simply

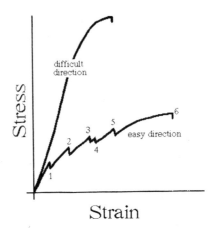

Figure 6.11 Stress–strain curves for a three-point bending test of crossed-lamellar material from a *Conus* shell with the crack starting to travel in the 'easy' direction and in the 'difficult' direction. Numbers refer to the stages of fracture shown in Fig. 6.12 (Currey & Kohn 1976).

from the fact that it is orientated differently, and that the crossed lamellar structure is weak if it is loaded such that the lamellae tend to be pulled apart layer by layer but strong if the crack tends to go between the laths in individual layers. This is easily resolved by tests with the stress applied at different orientations. It turns out that the toughness of the inner layer is due to its orientation: at the interface between the two layers the crack tip has to move off in two directions at right angles if it is to continue moving in the matrix phase. In fact it breaks across laths, effectively going from one lamella to the next, so the course of the crack is very tortuous. Figure 6.13 shows the sort of fracture surface which is produced. In these experiments, however, the first

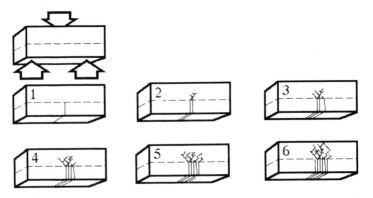

Figure 6.12 Three-point bending of crossed-lamellar shell in the 'easy' direction showing the state of propagation of cracks at various points numbered on the curve in Fig. 6.11 (Currey & Kohn 1976).

layer to be fractured has the crack travelling very easily between the lamellae rather like nacre in the weak direction. If the material is bent at right angles to this direction so that it starts off having to break in the difficult direction, the stress strain curve is very different (Fig. 6.11) giving a bending strength similar to that of nacre. The importance of the orientation of the lamellae relative to the advancing crack is also shown by some experiments with shell from *Conus striatus* which is thick enough to allow the preparation of specimens taken circumferentially. The crossed lamellar arrangement shows, as might be expected, other similarities with other composite systems as shown in Fig. 6.14: the strength is a function of orientation of the fibrous phase. Unfortunately more data are not available.

The strengths of other mollusc materials have hardly been investigated at all. Prismatic material, with its high organic content, would make a very interesting comparison with the other types with their much lower organic content. Homogeneous material seems to have more in common with the rubble used by the original builders to fill the walls of old (Norman or Romanesque) churches – it is extremely weak in tension since it consists of small lumps of material rather like expanded polystyrene on a small scale and has very little organic material. Like rubble it depends for its utility on its mass.

6.3 THE FUNCTIONAL DESIGN OF BONE

6.3.1 Where's the mineral?

Many books and articles have been written about bone. By far the best one covering mechanical properties is by Currey (1984b). He deals with a number of structural and design aspects of bone which are not covered in a strictly materials approach.

The mineral phase in bone is commonly referred to as hydroxyapatite, though informed opinion (Lowenstam & Weiner 1989) is that most 'hydroxyapatite' is in reality dahllite (carbonate apatite) or francolite (carbonate fluorapatite). Collagen seems to initiate the formation of crystals, although not all collagens will do so with equal facility. The implication is that not only does collagen provide some sort of template but also that some collagens provide better templates than others and can therefore direct calcification to some extent. The collagen presumably does this by providing an array of charged groups spaced in the same manner as the ions in the hydroxyapatite crystal lattice. This would provide a nucleating site on which crystals can grow epitaxially (Glimcher 1984). Certainly in bone the first crystals which form are precisely orientated with their crystallographic c-axis parallel to the long axis of the fibre and this orientation seems to be retained as calcification proceeds, as shown by X-ray diffraction and electron microscopy. These techniques show that the mineral crystals in mature bone are in the form of thin plates about 5.0 nm thick and about

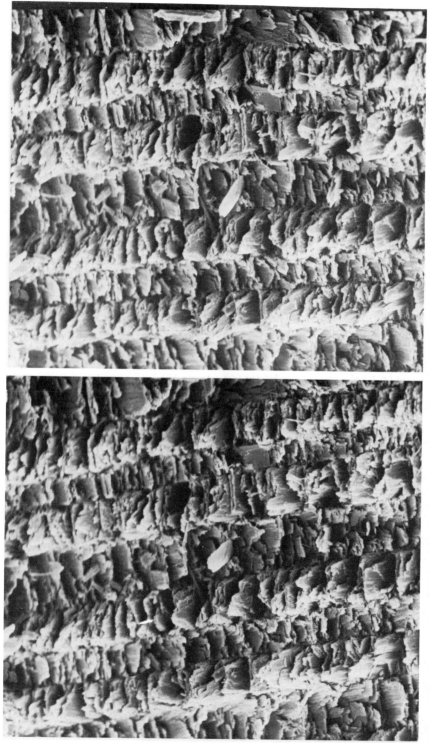

Figure 6.13 Stereo pairs of fracture surfaces through crossed-lamellar shell; A (facing page): broken at high strain rate; B, broken at low strain rate (courtesy of Professor John Currey). To be viewed with a stereo viewer.

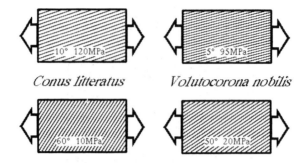

Conus litteratus   *Volutocorona nobilis*

Figure 6.14 The effect of orientation on the strength of crossed-lamellar shell material in tension (from data of Professor John Currey).

35 (standard deviation about 15) nm along the c-axis (about the length of the gap region in the quarter-stagger model of the collagen fibre) leading to a model of the type shown in Fig. 6.15 (Glimcher 1984).

Studies on turkey tendon (chosen because the collagen is highly orientated axially, giving a good diffraction pattern; because it has been well-studied by electron microscopy; and because the tendon is formed and functional before it is mineralized) show that the mineral is arranged with the same axial periodicity as the collagen and occupies the gap region. The lengths and widths of such crystals are of the order of 30 nm and 17 nm (Weiner & Traub 1986), although this width is almost an order of magnitude greater than that of a gap within a single collagen microfibril. So in turkey tendon (and by unsubstantiated analogy, bone) adjacent gaps must be in contact to allow crystals to overlap between them and to stretch

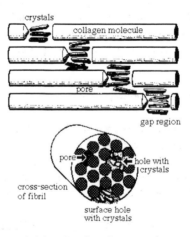

Figure 6.15 A possible model for the distribution of mineral crystals within bone collagen (Glimcher 1984).

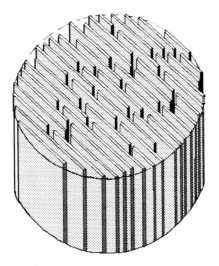

Figure 6.16 Distribution of mineral crystals in turkey-tendon collagen (Lowenstam & Weiner 1989).

across the entire collagen fibril (Fig. 6.16). The latest estimates of the size of crystals from mature bones (Weiner & Price 1986) gives the size as 53.5 nm × 28.5 nm from human cortical bone. This suggests that the Glimcher-type model is incorrect and that turkey tendon is a reasonable model system. It is possible that only about 75% of the hydroxyapatite in bones such as that of the cow's leg is sited in these gaps (Katz & Li 1973), but the margin of error on this number is quite large. All the same, the indication is that not all the bone mineral is intimately associated with the collagen. In addition, not all of it is crystalline, some is amorphous as judged by X-ray diffraction; bones of young vertebrates contain more amorphous material, becoming predominantly crystalline with maturity. As might be expected, this change in morphology is associated with changes in the mechanical properties.

6.3.2 Is it a 'simple' composite?

The major part of mature bone is thus made up of crystals of hydroxyapatite embedded intimately in a collagen matrix so it is possible to think of bone as a two-phase composite of crystals in a matrix. There is the complication that bone has holes in it to carry essential services to the bone and marrow cells, so some correction has to be made for porosity. Even so, there has been much agonizing over which phase might be the principal load-bearer. Wainwright *et al.* (1976) using Eq. 5.6 and the values given in Table 6.4 suggest that the minimum volume fraction of hydroxyapatite for it to act as a true reinforcement is about 0.35 which is almost exactly the volume fraction calculated by Katz

Table 6.4 Material properties of the major components of bone

Strength of hydroxyapatite	0.1	GPa
Stiffness of hydroxyapatite	130	GPa
Ultimate strain of hydroxyapatite	10^{-3}	
Stress in collagen at strain $= 10^{-3}$	1	MPa
Strength of collagen	50	MPa

& Li (1973) to be formed within the collagen fibrils. Since the volume fraction in cow leg bone is nearer 0.5, they deduce that the mineral phase is probably the main load bearer. But Eq. 5.6 has a number of assumptions built into it, mainly that the crystals are long enough to act as a fibrous reinforcer – i.e. that there is effective transfer of stress from collagen to crystal to collagen. It is possible that the hydroxyapatite crystals are long enough: nevertheless it is possible for the collagen to transmit high stresses, as in tendon, and that the collagen is the main load-bearer. The crystals would then be acting as filler particles stiffening the collagen by restricting its movement under stress. This would be similar to stretching a piece of rubber sheeting with or without lateral constraint. With lateral constraint the rubber will be more difficult to stretch. In three dimensions this will approach a fluid system and with a Poisson's ratio of less than 0.5 (say about 0.2) the reduction in volume on straining the bone would impose severe restrictions on the collagen, effectively stiffening it. This has been characterized as a 'straitjacket' effect (McCutchen 1975). For bone, Poisson's ratio is between 0.13 and 0.3, so the mechanism is a possible one, but will rely on orientation effects (see below). However, there seems to be no argument against simply considering bone as a composite filled with particles of no particular shape whatever, whereupon it conforms to the phenomenological model of Braem *et al.* (1987) and the pragmatically similar, though philosophically distinct, Hashin-Shtrickman model (Katz 1988). Both models are shown in Fig. 6.17. When it comes to considering bones covering a wide range of mineralization, the general trend of variation of modulus versus mineral content follows these latter two models (Fig. 6.18; Currey 1979; 1984a) although conversion of ash content to V_f of ceramic is a sum fraught with difficulty since the proportion of voids in bone can be very variable. Some estimates are shown in Fig. 6.19. The main line which is being followed by most workers in understanding the mechanical design of bone is that basically it is some sort of fibrous composite. Yet the mineral fraction of bone is predominantly platey, like nacre. It's only the collagen which gives a fibrous texture, and the collagen is only about 1% as stiff as the mineral. Judging by the fracture surface of antler, the fibre analogy can hold only if one views the fibre as being itself a composite reinforced with platelets. Exceptions may exist in some of the bones with a high V_f of mineral, such as dentin. Analytically, it seems, we are still confused, albeit on a somewhat higher plane.

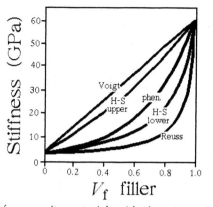

Figure 6.17 Stiffness of composite materials with changing volume fraction of filler. The Voigt & Reuss models (based on Fig. 5.3); the bounds of the Hashin-Shtrickman model (marked 'H-S upper' and 'H-S lower'; from Katz 1988) and a phenomenological model (marked 'phen'; from Braem *et al.* 1987).

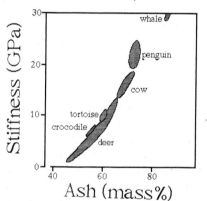

Figure 6.18 Variation of stiffness of bone, from a variety of sources, with varying ash content (Currey 1984a).

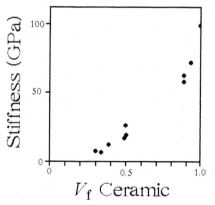

Figure 6.19 Volume fraction of mineral versus stiffness in bone, nacre, tooth and aragonite (plotted from a variety of sources).

6.3.3 Morphology of cortical bone

All the correlations between bone parameters so far discussed have neglected any structure of the bone at any level of organization higher than the mineral-impregnated collagen fibre. In fact bone has a complex hierarchical structure (Fig. 6.20); a large proportion of mammalian bone has a structure not unlike that of wood (Fig. 6.21). This obviously affects the mechanical performance of bone: the density and arrangement of packing of the mineralized fibres will

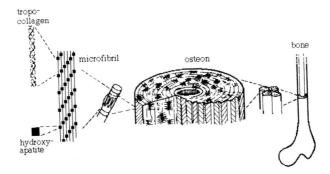

Figure 6.20 Hierarchical assembly of bone.

Figure 6.21 Wood-like osteons on a fracture surface in antler (Watkins 1987).

control the direction and magnitude of the stress which can be sustained and the way it is transmitted through the bone. Neutron diffraction studies of human scapula show that the c-axes of the hydroxyapatite crystals lie preferentially along the directions of pull of the attached muscles (Bacon *et al.* 1979; 1981). In areas where two muscles act in different directions the crystals are found to be orientated in two groups co-ordinated with the directions of pull of the two muscles (Figs 6.21, 6.22). There is some disagreement as to the morphology of bone, but the following is generally acceptable. The fibrils are arranged in two main ways: (a) in layers with a preferred orientation laid up to each other with the preferred directions varying much like the layers in insect cuticle or

(a)

(b)

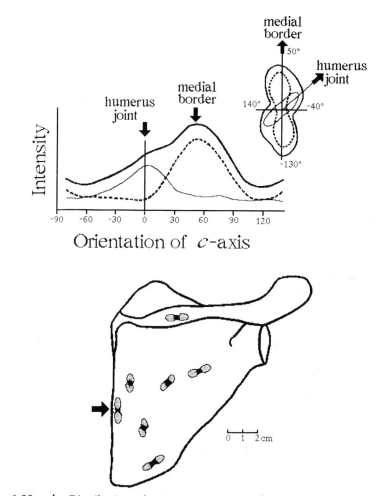

Figure 6.22 a+b Distribution of orientations of the c-axis of mineral crystals in part of the human scapula (arrowed). Orientations in other parts of the scapula are indicated (Bacon *et al.* 1979).

the secondary cell wall of wood cells (lamellar bone) and (b) more or less random (woven bone). Lamellar bone can be built into sheets in one plane (primary lamellar bone) or into cylinders like wood cells (Haversian bone; Figs 6.20 & 6.21) or mixed with woven bone (laminar bone). All types of bone structure are found in compact bone; the more open-textured bone found in the centre and ends of long bones and in vertebrae is constructed only of lamellar bone (primary lamellar or Haversian) and is known as cancellous bone.

Haversian bone is probably the commonest form of bone in the human skeleton and is widely found in other vertebrates. It is a replacement bone in that it develops after the bone has been largely calcified. Hence the cylindrical units comprising Haversian bone – the osteons – are strictly speaking secondary osteons. Down the middle of the osteon runs a blood vessel. The bone is laid down and maintained by osteocytes which sit amongst the concentric layers of fibres and send out radial processes (Fig. 6.20). The orientation of the fibres in successive is commonly considered to be discontinuous so that there is a sharp boundary in orientation between one layer and the next. Another model has been proposed in which the transition of orientations is more gradual, the general model conforming to the 'helicoidal' orientation model of Bouligand (Bouligand & Giraud-Guille 1985, Giraud-Guille 1988). Be that as it may, the complex of annular layers, cells and blood vessels is called the Haversian system. Adjacent systems are delimited by the cement line, a sheath of calcified mucopolysaccharide apparently lacking collagen. The osteon so formed is probably the basic mechanical unit of compact Haversian bone in the same way that the xylem cell is the basic mechanical unit of wood. It is different in that it can be branched but it is usually orientated along the long axis of the bone.

6.3.4 Morphology and mechanical properties

Following from the above arguments, one would expect that morphology would be second to mineral content in dictating stiffness. Currey measured stiffness, ultimate stress and the strain at yield in cow bones composed of differing amounts of laminar bone and Haversian bone. The effects on the mechanical properties of reconstruction and the falling ash content which accompanies it were inseparable. In other words the morphology of the bone seems to have little effect. By far the most important is the fact that laminar bone yields about 67% ash, Haversian bone 63%.

The stiffness of Haversian compact bone has been rather more closely investigated by Bonfield & Grynpas (1977). They measured the variation in Young's modulus at various orientations to the longitudinal axis of the bone (Fig. 6.23, circles and crosses) using an ultrasonic technique (the speed of sound through a material is a direct function of its stiffness). The stiffness does not fall off as quickly with change in orientation from the long axis as might be expected from fibre composite theory (steeply-falling line). But the composite model used has a single fibre orientation whereas the osteons of compact bone have more complex and varying fibre orientations. This experiment therefore confirms what has already become clear – that a simple-minded fibre composite model is inadequate for Haversian bone.

One way around the problem would be to test individual osteons (in this instance about 200 μm in diameter) in which the fibre orientation is known. Ascenzi (reviewed in Ascenzi & Bell 1972) did such experiments using osteons with three different types of fibre orientation. All osteons were tested in compression, a and b in tension as well (Table 6.5). So the orientation of fibres within the osteon is important.

Table 6.5 Variation in mechanical properties of osteons with orientation of the fibres

Orientation of fibres to long axis	Tension		Compression	
	Strength (MPa)	E (GPa)	Strength (MPa)	E (GPa)
~90°	116.5	11.9	112	6.45
~45°	95.9	5.59	137	7.54
~10°	—	—	167	9.49

From Ascenzi and Bonucci (1970)

A great weakness of Ascenzi's experiments was that he drilled his samples from the bone so that the osteons which he tested were not intact. This probably accounts for the rather low values for strength and stiffness compared with those for larger pieces of cow leg bone (Table 6.6), although it is possible for the morphology of bone to vary quite markedly over very short distances, so presumably the mechanical properties of small specimens could be locally variable as suggested by Table 6.5. Katz and his associates (reviewed in Katz 1980) found it possible to isolate single osteons. Tests on such osteons and groups of osteons (i.e. larger bits of bone) showed that the shear modulus of the osteons is greater the fewer osteons there are present. This is the opposite of what

Table 6.6 Comparison of two types of bone and whale bulla

Specimen	Porosity (%)	Stiffness (GPa)	V_f mineral	Toughness (kJ m^{-2})	Bend strength (MPa)
Foetal calf (a)	21.3	5.9	0.35	—	—
Antler (b)	—	7.4	—	6.19	179.4
Antler (wet) (c)	—	7.7	0.4	13	240
Antler (dry) (c)	—	12.8	—	2.14	—
Adult cow (a)	(19)	20.5	0.51	—	—
Adult cow (b)	—	13.5	—	1.71	246.7
Adult cow (c)	—	16	0.5	2.6	200
Whale bulla (a)	30	0.91	—	—	
Whale bulla (b)	31.3	—	0.2	33	

(a) Piekarsky (1973), (b) Currey (1979), (c) Watkins (1987)

Ascenzi's results suggested and indicates that the joints between the osteons have a much lower shear modulus than the osteon lamellae. This has been confirmed by creep experiments conducted on small specimens held in torsion over periods of several weeks (Lakes & Saha 1979). Specimens loaded like this show displacement of marker indentations due to movement at the cement line joint between the osteons. Such displacements account for nearly all the long-term deformation so that the joint between the osteons could be interpreted as a viscous interface. This interpretation is confirmed by the observation that dry osteons are Hookean up to a strain of 0.01 when tested singly; in a piece of bone containing ten osteons the Hookean region extends only to 0.0001. When wetted the osteons show Hookean behaviour up to 0.00001 even when single. Thus the joints seem to be important in controlling the balance between Hookean and elastic behaviour and may also bind water more effectively than the organic material in the osteon lamellae. And as with so many other biomaterials, water has a significant effect on viscoelastic behaviour.

If the osteon is the basic mechanical unit of secondary bone, in the same way that the xylem cell is the basic unit of wood, then a successful mechanical model might be possible using the osteon as the basic unit for the analysis. This analysis would have to take into account the orientation of fibres within the osteon as a primary level in the model and the cylindrical nature of the osteon and its relation to other osteons as a secondary level in the model. The fibre orientation can be modelled by 'simple' composite theory but the composite of osteons which

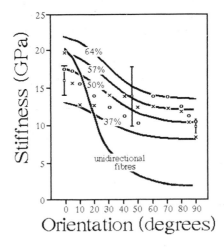

Figure 6.23 Comparison of a unidirectional fibrous composite model (marked 'unidirectional') with the more complex tubular composite model of Hashin & Rosen (1964) calculated for varying percentages of longitudinal orientation of the fibres (numbers; Katz 1980). The data points are for osteate bone measured at varying angles from the longitudinal axis of the shaft of the bone; circles, wet bone; crosses, dry bone (Bonfield & Grynpas 1977). The remaining three points (with standard deviations shown) are from Lugassy (1968).

is bone itself requires more sophisticated modelling, a successful version of which has been proposed by Katz (1980), based on a mathematical model for the effective elastic moduli of materials reinforced by parallel hollow circular fibres arranged in hexagonal array (Hashin & Rosen 1964). The exciting result is that the modulus of this type of composite varies far less with change in orientation to the longitudinal axis than does the modulus of a simple fibre composite, very much as found experimentally by Bonfield and Grynpas. The numbers on the lines in Fig. 6.23 (calculated by Katz) give the percentage of longitudinally orientated fibres in the individual model osteons. The stiffness of these osteons will vary with fibre orientation as shown by Ascenzi as well as by Katz and others. These values are then used in the calculation for the second level of complexity, giving rise to the lines drawn. The variation in the experimental results is probably due to such variables as the precise content of hydroxyapatite, the size of the osteons, the mean angle of the fibres, etc. Such variables can change over quite small distances so the fit of the model to data is moderately satisfactory. The variability is indicated by the three squares with associated standard deviation marked. These are from data by Lugassy (1968).

6.3.5 Viscoelasticity

The mechanical properties of bone are dependent on time (e.g. Fig. 6.24) and several creep and stress–relaxation experiments have been performed for a number of test morphologies. Lakes & Katz (1979) have derived a relaxation spectrum for human bone (Fig. 6.25) and, making the assumption that the intermediate maxima in the spectrum represent real events in the mechanics of the bone, have calculated the contributions of various dissipative processes to the overall loss processes (estimated by tan δ) measured in cortical bone

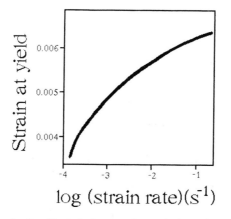

Figure 6.24 Viscoelastic effects in bone – the variation of tensile strain at yield with strain rate (Currey 1975).

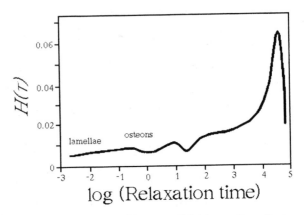

Figure 6.25 Relaxation spectrum of human tibial bone (to a first approximation) from relaxation and dynamic tests, with peaks assigned to lamellae and osteons (Lakes & Katz 1979).

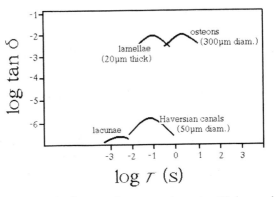

Figure 6.26 Loss angle peaks from various experiments with bone showing where the main relaxation times are (simplified from Lakes & Katz 1979).

(Fig. 6.26). This allows the relaxation processes in the spectrum to be associated with various morphological features. However, life is never that simple and bone has, in addition, been found to be thermorheologically complex: time-temperature superposition does not strictly apply so the variation in temperature which is used to increase the effective frequency range of the dynamic test equipment does not allow simple transformation of the data to a greater span of times in the manner of the WLF equation (Section 1.5).

6.3.6 Broken bones

Currey compared antler, cow leg bone and whale ear bone (Table 6.6; Currey 1979, Watkins 1987). His attention was first drawn to the toughness of antlers

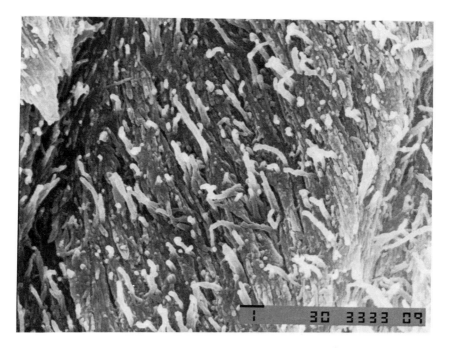

Figure 6.27 Fracture surface of antler showing collagen-mineral fibres separated (Watkins 1987).

by the uses to which Man has put them – notably as pick-axes and a material from which to make long-lasting combs. Both these uses demand a tough material even though, as an item of display, antlers need to be made of nothing more sophisticated than, as Currey says, waterproof cardboard. But when stags fight the clash of antlers reverberates for miles around. Clearly the antlers are absorbing much impact energy and transferring it to the muscles of the stag's neck where most of the energy of fighting is absorbed. So the antlers must be tough if their owner is to win the confrontation. This is shown in impact fracture tests where antler breaks into only two pieces, suggesting control of the crack path, whereas cow leg bone shatters in an almost glassy manner. Notice that the antler is the least mineralized and thus the least stiff of bones, suggesting that it can reach a relatively high strain. In fact antler seems to have confined the mineral phase to within the collagen fibrils; the fracture surface shows fibres less than 200 nm in diameter (Fig. 6.27) which is only slightly larger than the size of fibre reported for mineralizing turkey tendon (Weiner & Traub 1986) and the amount of mineral corresponds to the space available within the fibre (Katz & Li 1973). It seems that at this low level of mineralization there is no cementing mineral around the fibres so that they remain free and unadhered. This is also indicated by the tensile stress–strain curve of antler – it shows significant plastic yielding, presumably due to the fibres pulling out. However,

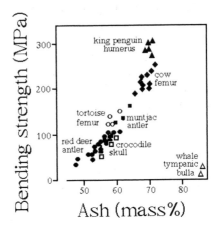

Figure 6.28 Bending strength of bones from various sources, showing variation with ash content (Currey 1984a).

the low level of mineralization also means that antler bone is not as strong as bones with higher mineral (= ash) content (Fig. 6.28; Currey 1984a). Cow leg bone has an intermediate stiffness and degree of mineralization. Although it is not so tough as antler it is stronger and can withstand higher forces: it does not deflect so much under load. Indeed it does not need to be as tough as antler since in life it is covered in flesh and skin which will cushion shocks to a great extent. The ear bone or tympanic bulla of the whale is never submitted to such shocks. Its function requires that it should be have high inertia, so it is very massive and has more mineral in it than any other bone. So although it is very stiff (Fig. 6.18) it becomes weak and brittle and has a smooth fracture surface not unlike a glass.

It is symptomatic of the sort of compromise which biological materials have to achieve that cracks probably commonly start from the holes and tubes which bring the bone its lifeblood. If bone is broken slowly it is possible to watch cracks starting and progressing through the bone. In one study, cracks started to form at about 60% of the failure load. In one instance a crack 230 μm long was seen moving from a small canal, about 25 μm in diameter. In most instances the cracks moved between the lamellae of the osteons starting from the cavities occupied by the osteocytes. This latter mode seems to be more common in fracture at low strain rates and the cracks are then deflected around the osteons between the lamellae. The cracks can be seen to originate more commonly at blood vessels, which are anything from 20 to 100 μm in diameter. This conclusion was reached independently by Bonfield (Bonfield & Datta 1976) using an analysis based on Griffith's ideas of fracture (Section 1.6) rather than direct observation. He calculated an 'intrinsic crack length' from which further cracks could start, and found it was about 340 μm, which is nearer to the size of the blood vessels than of the osteocytes (3–5 μm).

The fact that bone is viscoelastic suggests that the way bone breaks, and its toughness, should be a function of strain rate. It should also be dependent on the degree of viscoelasticity – i.e. the amount of plasticizer (= water) in the collagenous matrix – in the same way that nacre is (Table 6.6; Watkins 1987). At high strain rates, achieved by impact loading of various sorts (including the use of explosives) cracking does not follow any particular path through cow leg bone. The fracture surface is as one might expect: at low strain rates it is very convoluted showing a very rough surface. Osteons are pulled out whole very much like fibre pull-out in artificial composites. Fragments of osteons can also be seen, all indicating that the crack did not propagate perpendicularly through all the bone constituents. The fracture surface of cow leg bone produced by failure at high strain rates shows no evidence of pull-out and the surface is much smoother – like that of the tympanic bulla of the whale. Evidently these differences are a result of the viscoelastic response of the bone components to different strain rates. In a study on cow leg bone Robertson and Smith (1978) found that at high strain rates the stress–strain curve was linear with a brittle failure typical of a glass; at low strain rates, with ample time given for plastic relaxation to occur, the stress–strain curve was not linear, the bone yielding and supporting load to higher strains. The transition between these two modes of fracture occurs at a strain rate of around $2.5 \times 10^3 \, s^{-1}$ (i.e. it would take 4 s to reach a strain of 0.01).

It seems reasonable to think that this strain rate dependence will be shown in the work required to fracture the bone, but opinion seems to be divided. Mostly it seems that the work of fracture is related to strain rate but it seems likely that, as Currey has pointed out, the relationship between strain rate and various other mechanical parameters depends strongly on the structure of the bone – whether it is osteate, lamellate or intermediate in some way or another. These structures can produce very different effects. It is also important that the sample should have time to pay off all its stored strain energy into the processes of fracture, otherwise the energy may be overestimated. Behiri & Bonfield (1982) reported that the toughness of cow leg bone increases as the velocity at which the crack tip travels also increases, up to a value of about $1.2 \, mm \, s^{-1}$, going from 0.63 to $2.88 \, kJ \, m^{-2}$. At this velocity the propagation of the crack becomes unstable, the fracture surface changes from being rough to glassy, and the fracture toughness drops to about $0.2 \, kJ \, m^{-2}$. On balance, it seems that osteate bone is weaker at higher strain rates, but again such factors as the size of the osteons can affect the result – larger osteons seem more prone to brittle fracture, presumably because the path of the crack is not deflected so often.

The ultimate effect of the fracture behaviour of bone is to produce a material which is much stronger in tension than would be expected from a predominantly ceramic material. Simple experiment (Table 6.7) confirms what observation has already suggested – the organic phase is the important factor in increasing strength.

Table 6.7 The importance of the organic phase to the mechanical properties of bone

	Tension		Compression	
	Strength (MPa)	E (GPa)	Strength (MPa)	E (GPa)
Normal	130	17	150	9
Organic phase removed	6	17	40	7.2

6.3.7 Bone as a cellular solid

Even though not explicitly stated, the discussion so far has concerned 'solid' or cortical bone as found on the outside of an intact bone, composing the walls of the main shaft of limb bones. But the inner structures of many of these bones, especially where extra bracing is needed for off-axis loads (e.g. at the joints), can be very extensive with networks of trabeculae. The most familiar example is the femoral head, classically figured in D'Arcy Thompson's *On Growth and Form* (1917) and described by Whitehouse & Dyson (1974), but one of the nicest examples is the palmate antler of the moose. This is up to 20×50 cm and 4 cm thick with relatively dense cortical bone grading into an open meshwork of cancellous bone in the middle. Gibson & Ashby (1988) call this a sandwich structure and group it with skis and helicopter rotor blades. The compressive stiffness and strength of cancellous bone conform closely to the general models for cellular solids (Gibson 1984), varying with the square or cube of the relative density depending on whether the individual cells are open (i.e. assemblages of struts) or closed (i.e. plate structures).

6.4 TEETH

In the other major ceramic materials of vertebrates – tooth enamel and dentin – tensile forces of any appreciable magnitude occur much less frequently. The major requirements of a tooth are that it should have a hard surface for cutting and grinding (whatever they might be in physical terms) and that it should be durable (= tough) enough to last for the lifetime of the animal. The human tooth (Fig. 6.29) is a fairly basic type with dentin core and enamel capping. Dentine and bone are similar in overall composition (Table 6.8) but enamel contains much more mineral.

Dentin is similar, in the main, to bone. It is more uniformly constructed than bone but the crystals are probably much thinner – about $2 \times 50 \times 25$ nm. The dentin is permeated by tubules (Fig. 6.29) which are surrounded by a highly calcified zone (the peritubular dentin) and sit in a matrix of randomly orientated crystals in mucopolysaccharide and collagen, the latter orientated in planes parallel to the surface of the dentin.

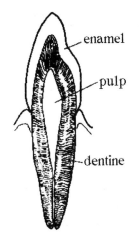

Figure 6.29 A human incisor.

Enamel has a much more complex structure. In newborn mammals, the spaghetti-like crystals of dahllite (carbonate apatite) are at least 100 μm long but may be only 50 nm in diameter (Daculsi *et al.* 1984) so that each crystal extends from the surface of the tooth to the junction with the dentin. Such a high aspect ratio allows these crystals to assume a variety of curvatures along their length, which may account for previous reports of a variety of orientations of smaller crystals within the large components into which these crystals are organized. These components are various sorts of rods and sheets, depending on the animal. The crystals can branch and fuse, leading to the formation of pyramidal shapes with the wide base projecting towards the dentine layer. In mammals the arrangement of these rods and sheets can be complex and various, in some instances giving a completely three-dimensional array of rods. The arrangements of crystals in lower vertebrates are different again (Lowenstam & Weiner 1989). Two classes of protein are present in enamel. The first are the acidic enamelins which are covalently associated with polysaccharides and tend to adopt β-sheet conformation. They are associated with the control of shape of the enamel crystals. The other proteins are hydrophobic amelogenins.

Together the dentin and enamel of the tooth are subjected to loads of 20 MPa, 3000 times a day (less if you are on a diet). Even so, fracture of sound teeth

Table 6.8 Percentage composition of bone compared with those of dentine and enamel

	Bone	Dentine	Enamel
Mineral	66	70	95
Organic	24	20	0.5
Water	10	10	4.5i

From Eastoe (1971)

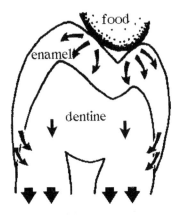

Figure 6.30 Forces on a molar analysed by the method of finite elements (Waters 1980).

is very rare. This seems to be partly due to the hardness and stiffness of enamel and partly to the toughness and relative compliance of dentin. The mechanical properties of dentin and enamel has been measured using a variety of tests (Table 6.9; Waters 1980, Fox 1980, Staines *et al.* 1981). The higher toughness of dentin parallel to the tubules is probably related to the orientation of the collagen – the crack in this plane has to cross the collagen layers. Similarly the greater toughness of enamel is probably related both to the presence of weak interfaces between the prisms and the path of the crack through the prisms which will be retarded by the highly fibrous morphology of the crystals. Fox (1981) suggested that additional toughening comes from the movement of fluid through the structure of the enamel, dissipating energy in doing so. Parallel to the prisms the toughness is what one might expect of a normal ceramic material. The functioning of the entire tooth has been analysed by finite element analysis (a particularly diabolical way of using a computer which considers the elastic response of large numbers of small elements of the material), and Fig. 6.30 shows the results of such an analysis of a biting molar tooth. The arrows show the directions of the stress trajectories.

6.5 EGG SHELL

If the tooth is a difficult object to analyse and test, the egg shell of birds must be nearly impossible. Moreover since materials scientists have by and large kept clear of mechanical tests on egg shells, thus demonstrating a clear understanding of the difficulties of such tests, the literature on the strength and other properties of the shell is dominated by half-boiled notions, mostly from biologists and agriculturalists. Rehkugler has estimated the stiffness, using circular test specimens cut from across the shell, as 10–20 GPa which is about a third the

Table 6.9 Comparison of properties of dentine and enamel

	Compressive strength (MPa)	Stiffness (GPa)	Vicker's hardness number	Orientation of crack plane	W_f (J m^{-2})
Dentine	300	12	70	perp. to tubules	270
				parallel to tubules	550
Enamel	200	40 to 50	300 +	perp. to prisms	200
			parallel to prisms	13	
	330	74 to 84			

From Waters (1980)

stiffness of limestone. But this overlooks aspects of the morphology of the shell which could affect the analysis of the results. In fact it seems that the indentation hardness varies through the thickness of the shell of the domestic hen's egg, being about 170 kg mm^{-2} on the inside and outside, dropping to about 125 kg mm^{-2} in the centre (Tung *et al.* 1968). The eggshell is made of calcite, about 96–98% by volume, the rest being hydrated organic material, but the morphology of the shell material (Fig. 6.31) is complex and mechanically enigmatic. The organic material is spread throughout the shell material so that if the carbonate is dissolved a 'ghost' of organic material remains. The fracture surface of the mineral varies throughout the thickness of the shell: in the cone layer the surface is like that of prismatic mollusc shell; in the main thickness of the shell the fracture surface appears much less ordered although the orientation of the calcite crystals is in fact very highly ordered (Silyn-Roberts & Sharp, 1985a,b). It seems possible that the fracture planes are defined by the distribution of organic material. The tips of the cones are connected by a membrane which is highly fibrous. The eggshell as a whole has a complex shape and is brittle. It is therefore very difficult to prepare a sample for mechanical testing. Any strip or beam cut from the shell will have a highly complex shape and may well contain microcracks as a result of preparing the specimen. Most shells are too thin for a rectangular cross-section to be machined; in addition it is difficult to decide what part of the shell section is actually taking stress. As an illustration of this, Bond *et al.* (1986) suggested that when the egg is loaded from the outside (typifying the protection function of the shell) the membrane, acting in tension, may be shifting the neutral axis of the section effectively inwards thus throwing more of the outer part of the shell into compression (Fig. 6.32) in a similar manner to the model proposed for insect cuticle by Ker (Fig. 5.19). This would relieve the cone layer from what little tensile stress it might see and may explain why the mineral in this layer is so crystalline and so, probably, weak. It is, indeed, doubtful whether this layer would see any stress at all in the plane of the shell since the cones are not connected laterally. It is thus conceivable that the function of the cones is to hold the tensile membrane away from the neutral axis. Moreover it is likely that the membrane

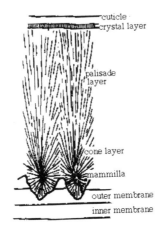

Figure 6.31 Diagrammatic section of an egg shell (about 200 μm thick).

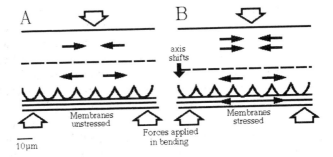

Figure 6.32 The shell membrane and the shell itself acting as a complex beam system under loading (see text).

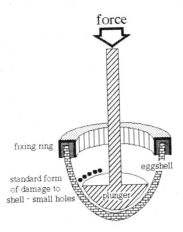

Figure 6.33 A mechanical test designed to simulate hatching.

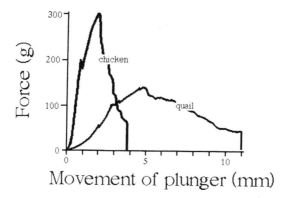

Figure 6.34 Some results from the test shown in Fig. 6.33 (Bond *et al.* 1986).

is prestressed in tension since it dehydrates soon after the egg is laid and so probably shrinks. This would shift the neutral axis still further towards the tensile side and so effectively allow the shell to take even higher compressive stresses. This analysis is hypothetical but illustrates the possible complexity of the shell structure and therefore the difficulty of interpreting any mechanical tests. However, such a structural analysis goes well with the old story that it is more difficult for the hungry breakfaster to get into his egg than for the chick to escape from it. This is often attributed to the wedge-shape of the shell units associated with each cone and with the shape of the egg as a sort of three-dimensional arch made entirely of keystones. But it is also an oversimplification since the chick, before attempting to hatch, has very wisely eroded away about half the thickness of the shell and incorporated the calcium within its own tissues. The chick in fact escapes from the egg by chipping away a circle at one end of the egg and pushing on the disc formed so that the shell will tear along the dotted line. Bond *et al.* (1986) have simulated this in a mechanical test (Fig. 6.33) and shown that the eggs of the quail and pigeon are made of much tougher material than those of the hen or the duck (Fig. 6.34) and that this is reflected in the hatching behaviour of the chick. Hen and duck chicks need peck only part way around the shell before pushing off the disc to emerge; quail and pigeon chicks have to peck right around the shell before pushing. Thus, as any gourmet will tell you, crack propagation is easier in duck and hen eggs than it is in quail or pigeon eggs. The reason for this is obscure. Bond considers that it is most likely due to small differences in the amount and composition of the organic material. There seems to be no difference in the appearance of the broken surfaces of the two classes of egg shell.

6.6 ECHINODERMS

Some of the most intriguing ceramics are produced by the echinoderms. Their skeletons are composed mainly of magnesium-bearing calcite containing less than 1% of organic material. Individual components (plates from the test, spines, teeth, larval spicules etc.) behave as single crystals when examined by X-ray diffraction or in polarized light. But when they fracture they behave in a number of ways. Wet pieces of the test of *Echinaster spinulosus* were loaded and fractured. If they were fractured immediately, the fracture surface was smooth and conchoidal, typical of a brittle (though amorphous or glassy) material. But if the pieces were kept loaded for some time before they fractured (i.e. allowed to stress-relax) the fracture surface was rough with lots of small needles sticking up about 0.36 μm above the surface, arranged as if they represent the broken ends of concentric lamellae (O'Neill 1981). The inference is that the needles represent components which are surrounded by a viscoelastic material which deforms during the stress relaxation, giving weak interfaces. Thus the calcite must be polycrystalline. The puzzle is – how do the crystals (if they exist as separate entities) grow in such perfect register? One way to investigate the mechanism is to grow the crystals yourself. Berman and friends (1988) did just this, growing calcite crystals in the presence of matrix proteins both from *Mytilus* shell and a sea urchin test. The calcite grown with urchin matrix protein was tougher and broke in a glassy manner rather than cleaving like a crystal. The *Mytilus* protein also affected the fracture behaviour, but the resulting crystals crumbled instead of cleaving. The inference is that echinoderms are making a 'molecular composite' with the protein being laid down on the growing surface of the crystal and continually being overgrown by further calcite. In doing so it may well be providing the surfaces along which O'Neill's needles separated from each other. Thus echinoderms manage to have the best of both worlds – the strength and stiffness of a ceramic, but without the brittleness.

The stiffness of echinoderm larval spicules has been measured as around 40 GPa (Emlet 1982); 69.4 GPa and 52.1 GPa for dry and fresh spines, respectively, of adult *Diadema setosum* (Burkhardt 1983); 74 GPa (spines) and 9.7 GPa (test) by Currey (Wainwright *et al.* 1976). This is about half to a third the modulus of the pure calcite, leading to successful modelling of stiffness by the Reuss model (Emlet 1982).

The five teeth of the Aristotle's lantern have been shown to be complex ceramic structures with varying amounts of organic matrix, although the crystalline structure still appears (when studied by X-ray diffraction and polarized light) considerably simpler than that revealed by fracture. The hardest part (the stone part) which is the working tip, is an orientated composite of dolomite (it is not known if fibre and matrix contain different Ca:Mg ratios, which might be expected) with fibres 2 μm in diameter and a critical transfer length calculated as 16–18 μm. The hardness can be as high as 168.5 kg mm^{-2} (in *Clypeaster rosaceus*; Maerkel *et al.* 1971) although this is nowhere near as hard as *Patella* radular teeth (this fact alone suggests that the values recorded for *Patella* radular teeth are at least an order of magnitude too great – see Chapter 5). Clearly there is also some confusion with the measurements of hardness of echinoderm teeth, since a figure for hardness as high as 300 is quoted for the stone part of *Sphaerechinus granularis* (Maerkel 1974). Perhaps these teeth had been stored in alcohol. The urchin teeth have reduced hardness behind the leading edge of the working tip giving a self-sharpening system similar to that of rodent incisors and locust mandibles. Using different starting materials but requiring a similar function, these animals have arrived at a very similar structure for their teeth.

CHAPTER 7

Biomimetic and Intelligent Materials

If skin, tendon, hoof, horn, bone and nacre are such wonderful pieces of materials engineering, why don't we try to make them ourselves, or at least use some of the ideas contained within them to our own advantage? As our understanding of materials in general, and biological materials in particular, has been advancing, so has our ability for biomimicry. But the ability to do something has to be driven by desire or necessity. Why should engineers be interested in biological materials? One only has to look to the history of technology, starting from the first use of tools, to see that revolution and progress are driven by the new possibilities which new materials create. The initial and most obvious effects are frequently expressed as advances in the technology of weapons. So bronze yields to iron which in turn yields to steel.

On a broader level, man has progressed from using natural materials (for instance antler combs and pickaxes, stone tools) through the modification of natural materials by chemical means (for example the smelting of metals, the cross-linking of rubbers) and the primary synthesis of new materials such as plastics to the ability to design new materials starting from the atomic and molecular level. At each stage the properties of the materials and structures which have been produced have represented a significant advance in quality of performance, improving on such parameters as stiffness, strength, durability, ability to withstand extremes of temperature, fatigue resistance, etc. But with sophistication in production and design of a material comes sophistication in use, so that materials are required to function better at whatever particular task they are used for. So we find glass fibre reinforced plastic (GFRP) is used in different ways depending on whether it is part of a high performance sailplane, a car body or a chair. And if you look at the way a modern sailplane is made, you find that GFRP is used in many different ways around the various bits (wings, fuselage, tail surfaces, etc), producing a light and 'efficient' structure with a high performance. But this is exactly the way evolution has dealt with the materials from which animals and plants are constructed. The tendency must be to minimize the energy input, whether that be in the form of synthesizing materials or controlling their shape, and to maximize the performance. So animals and plants are doing just the sort of optimization that the modern engineer wants to incorporate into his new efficient designs. And a large part of this optimization involves not just choosing the right material for the job, but tailoring the material for a particular application. Think only of the number of different ways insect cuticle can be put together to provide materials which are stiff or flexible or rubbery or stretchy or opaque or translucent.

So the impetus for the production of materials and structures using a biomimetic approach comes from the realization that increased efficiency, specialization and optimization are common driving forces in engineering and biology (Birchall 1989). The corollary remains as it always has done. The increase in understanding of physics and materials which the physicists, chemists and engineers can bring with their studies on simpler materials brings insight to the study of biological materials and thus the organisms which produce those materials. So just as composite theory helps us to understand insect cuticle and bone, so one must expect that research into methods for the production of biomimetic materials will provide clues to the cellular processes involved in the production of the skeleton. And this will inevitably feed back into improved synthetic control of biomimetics. It is a true symbiosis. As many people have claimed to say, biology is the basic science, physics is just a special case!

New materials are initially developed for areas of use in which money is not limiting. These mostly turn out to be defence and medicine (most people will pay anything for the promise of a long life) and leisure activities (which may explain why the Olympic Games are the financial as well as the moral equivalent of war). One of the goals in medical research has been the development of replacement tissues. Frequently these are almost obscene in their naive boldness. Imagine replacing an artery with its complex elastic anisotropy with a piece of plastic tubing. Imagine also replacing a hip joint with pieces of metal and plastic. This turns out to be rather more successful, but there are still problems in matching the mechanical properties of the metal with the mechanical properties of the bone. Where the two meet there will be, if you are not very careful, a stress concentration which will lead to the failure of the bone. So do we know enough about bone to be able to make a satisfactory replacement, temporary or permanent? Two areas of knowledge are needed: will the replacement material do the job as well as the original material did, and how can its properties properly be matched to those of the remaining bone? Materials scientists, used to making materials more or less by mixing a bucketful of this with a bucketful of that, come up with ideas such as hydroxyapatite in plastic. In fact this is a rather successful approach (Bonfield 1987), but the two parameters of stiffness and toughness can still not be matched simultaneously in an artificial material in the way which bone can.

Since stiffness is widely perceived to be an important attribute (probably because our maths lacks the subtlety to cope with the higher strains of more compliant materials), many people are trying to copy biological ceramics. The main problems (Section 6.1) seem to be producing crystals which are small enough (since the surface area of a fracture surface is probably an important factor in controlling toughness) and well enough bonded to the matrix material. It is possible to produce pure ceramic materials made of small particles which are adequately stuck together by the 'sol-gel' process (Jones 1988), although the toughness of such materials is orders of magnitude less than that of a biological ceramic composite. There seem to be two main alternatives at present: either grow small crystals under very controlled conditions (Mann *et al.* 1988), when it is still necessary to embed them in a matrix of some sort, or follow biological systems even more closely and form ceramic particles within a matrix phase by *in situ* precipitation or crystallization (Mann & Calvert 1987, Calvert & Mann 1988, Calvert 1989). The main problem at present seems to be control of orientation of the ceramic phase, probably because the matrix material is not directing the formation of the ceramic phase sufficiently closely. Only with adequate control over orientation can high ceramic volume fractions be achieved. But man is ever impatient. Biomimetic processes always take minutes instead of the months which biological processes take. Part of the secret of biological processes is that growth always occurs on a free edge or face. Biomimetic materials which are made in bulk by processes such as precipitation within a gel run into two basic problems. First, the soluble ceramic precursor necessarily reduces in volume when it is converted to a solid phase; second, the gel tends to shrink uncontrollably and crack when dehydrated. These two effects lead to a lower volume fraction of ceramic in a less coherent matrix. It is not difficult to think of a number of mechanisms which might overcome these problems. One is to add polymer to the growing crystals so that they incorporate the matrix as they grow. This has been shown to be successful with calcite crystals growing as a model of the echinoderm skeleton (Berman *et al.* 1988) and is not too different from the approach taken by Birchall's group producing macro-defect

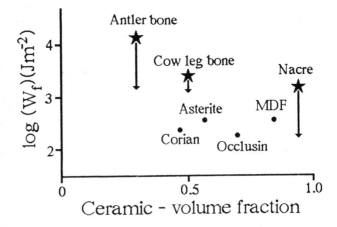

Figure 7.1 Toughness of natural (star) and artificial (dot) ceramic composites. The values for the natural composites are for the wet (plasticized) materials. When dried their toughness drops, as indicated by the arrows.

free (MDF) cement (Roy 1987). In this process a water-soluble polymer such as polyacrylamide is added as a rheological aid to permit cement to be mixed with very small amounts of water. By using particles of varying size distributions so that there are no holes left in the material, and applying a pressure to drive out any air pockets, it is possible to end up with a material which is 80% ceramic and quite tough. But still nowhere hear as tough as a biological ceramic would be (Fig. 7.1). An alternative approach using monodisperse silica particles in methacrylate monomer, subsequently polymerized, produced materials with good performance (Jin *et al.* 1989), but the use of spherical particles limited the volume fraction of ceramic to about 60% (the theoretical maximum is about 75%). This may be the reason why platelets are so common in biological ceramics – they can be packed together much more closely.

Biomimetic materials can also provide new technological possibilities and advantages. The drive for novel adhesives based on novel proteins used by mussels and barnacles is driven by the current lack of adhesives which will cure in a wet environment, the poisonous nature of solvent systems necessary for currently available adhesives, resistance to moisture penetrating into the adhesive layer, the ability of such adhesives to cope with dirty surfaces and the need for medical adhesives. However, there is no need to be limited to extant bioadhesives. Protein glues such as casein and gelatin have been used for hundreds of years. Such glues take their strength from the covalent bonds of the protein backbone and are routinely as strong as any modern artificial adhesives. They can be enhanced by incorporating an epoxy group (Kaleem *et al.* 1987) or phenolics (thus mimicking the natural system), urethanes, methanes and polyesters (Barrett 1984). In this way their inherent 'stickiness' can be exploited while circumventing their associated tendency to absorb water and either dissolve or rot.

Soft materials are also of interest for biomimicry, for medical use (artificial skin and artery) and use in robotics. Obviously developments in robotics can be fed back into orthotics and prosthetics. But in general, medical use seems to regard physiological matching as more important than physical compatibility. You can always tell someone to slow down until he is either better or dead, thus giving the implant a better chance at coping. Robotics is more interesting from the materials point of view since the very idea of using a robot is that it should *not* have to slow down! The compressive properties of skin can be mimicked to provide tactile sensors (de Rossi *et al.* 1986, de Rossi 1989), although the materials used are mechanical mimics only, and biology is useful only in that it provides the properties which are to be mimicked, rather than the means of doing so. This latter approach has been taken in the mimicry of tendon collagen. Using crimped fibres of Kevlar in a matrix of poly(methyl acrylate), Pradas and Calleja (1990) have made an artificial tendon material with a J-shaped stress–strain curve which has similar damping and stress-relaxation properties to real tendon. Here, however, the aim is to make a prosthetic material rather than develop a new material which can be used in non-biological devices. Perhaps that time will come.

Biomimicry is leading not only towards materials which have a better mechanical performance but towards materials that can react in some way to their environment. This concept has been called 'intelligence' in a material, and such materials (and associated structures) are also called 'smart'. So the response of skin to damage (Section 4.6) could be called 'intelligent' in that further damage is thereby reduced. However, a biological material can also be remodelled or structurally modify itself in response to environmental stresses. So it is able to redesign itself in response to the way it is being used. The transduction between strain and cellular response is usually thought to occur as a result of piezoelectric properties of a component of the material. So if the material is being strained in uneven manner the piezoelectric effect will cause a small electric potential to be generated which can be detected by the cells responsible for synthesizing the material. For example, collagen is piezoelectric (though only weakly so when wet) so the osteocytes may be able to detect whether the bone is being bent or being loaded off-centre as might happen if a broken bone does not join properly in line. Another possible mechanism for transduction is streaming potential, which can occur where a liquid capable of carrying ions is free to flow, causing a potential difference between the upstream and downstream ends of the flow. The way in which these effects might work to tell osteocytes how to remodel bone is discussed by Currey (1984). Mattheck (1989) and Mattheck *et al.* (1989) have analysed a number of biological shapes (mainly trees) and shown that they conform to two general design principles – they have sufficient strength for all relevant instances of loading and they are of minimum weight for the job they do. Both Mattheck and Topping (1989) discuss the way in which finite element analysis can be adapted to generate such optimized shapes. In many instances this involves subtle changes to the radiusing of joints. The net effect of such mechanisms is to cause *more* material to be deposited in areas of greater load. But the very opposite occurs in a passive inorganic system – a loaded spring will corrode much more quickly than a non-loaded one. Such stress-corrosion always guarantees that the most highly stressed joints are the first ones to break. A smart or intelligent material would act like bone and wood and cause more material to be deposited in a highly stressed area. This would allow the development of self-designing structures. All you do (for instance) is make a wire frame representation of the structure you wish to build, load it in the way you expect it to be loaded when it's being used, and allow it to accrete material where the stresses are greatest until the stresses are evenly distributed. Something along these lines is being tried to produce self-building structures in the sea (Bubner *et al.* 1988, Hilbertz 1988). Calcium salts are made to precipitate on various materials, charged negatively from a small battery. If the distribution of charge could be controlled by strain detectors sensing local loads, then a self-designing structure would be possible. This is one of the goals of certain modern architects.

The intelligent material will continue to modify its structure throughout its useful life, so that although the applied loads and their directions may change during this time, the material will be able to respond by changing its shape. It will be able to do this on a longer time scale and (for instance) increase or decrease the size of various components, or it will be able to change the shape of its structure on a short time scale, so that instead of supporting loads from various directions by increasing in mass and stability (the solution chosen by the builders of cathedrals), it will be able to move and adapt its shape to that of the web of forces within which it exists. This, of course, is what animals and plants do all the time, and it allows loads to be supported much more cheaply. Such structures and materials may take some getting used to: imagine instead of a cathedral-like structure a house supported in a web of tensile elements which is able to support itself and respond to external forces. It would lean into the wind! Buckminster Fuller used to play around with ideas like this.

Obviously such materials and structures need a variety of control systems at different levels, both local and global. The local control can be built into the material; the global system requires an integrated system of sensors. This takes 'intelligent' materials into the realm of smart structures.

CHAPTER 8

Experimenting with Biomaterials

8.1 INTRODUCTION

Biomechanics is sufficiently primitive as a science for significant experiments to be done with relatively little equipment. No doubt this state of affairs will change as all the easy experiments are done, but for the moment it takes only a clamp, a few weights and a piece of string to make useful predictions about the effect of wave action on a piece of seaweed. In Reading we have been running a short summer course of biomechanics experiments for some 15 years and in that time have accumulated a number of ideas for projects which can be performed with relatively little equipment and yield significant results within about 20 man-days. What follows is based on that experience.

A biomechanics laboratory can be a simple affair with little more than a clamp, a few weights and a ruler. But ideally one would also have some sort of universal testing machine capable of loads up to 5 tons or so. You will never probably need such high loads, but the machine needs to be stiff enough not to deflect under the load you *will* use. The machine must have interchangeable load cells, capable of tension and compression, going from 10 N up to 10 kN in four stages, and pneumatic clamps rather than screw or wedge ones. The crosshead can be moved by a screw thread or (preferably) hydraulics. Output can be to a chart recorder (for preference) and/or computer. Such a machine can perform tests in tension, compression and bending. A torsional pendulum is also useful, especially for small delicate specimens and for measuring shear stiffness at low strains. Most important is a box full of general rubbish (metal strips, nuts, bolts) from which machines and adaptors can be built, together with a selection of strain measuring devices and load cells. Biological materials and structures rarely come in the shapes and sizes which the machines were originally built for. You need to be able to make your own *ad hoc* machinery.

You also need some reference books for general analysis and some ideas as to the limitations of any particular test configuration. The following is a minimal library: Roark & Young (1975), Ferry (1970), Stephens (1970), Timoshenko & Goodier (1970), Atkins & Mai (1985), Piggott (1980). These are mostly in the 'cook book' category as far as most biologists go, so it is as well to have access

210

to a sympathetic engineer as well who can help you through all the assumptions involved in deriving the mathematical analysis. It may also be useful to try some of the experiments initially with simpler materials such as metals and plastics, which can also be obtained in simpler shapes and larger sizes. Vincent (1978) describes a number of these experiments, including some ideas on model 'biological' materials.

The experiments which follow have been chosen mainly because they are relatively simple to perform and give clear-cut results. In general it is easier to design a more difficult experiment. The experiments are arranged in terms of the techniques used rather than the materials or organisms.

8.2 TENSILE TESTS

8.2.1 Stiffness and strength

These are probably the most common and easily appreciated type of test. There are some provisos which attach to all tensile tests. The material must be free to change shape, so that Poisson contractions orthogonal to the direction of extension must be freely allowed or the material will appear too stiff. In practice this means that the constraint at the ends where the material is clamped must make an insignificant contribution to the overall properties, or that the extension of the test piece should be measured independently of the movement of the machine crosshead. If (as you should be) you are a little paranoid about this, then it is possible to measure the 'end effect'. Using samples of different length, plot length against (extension × compliance). The negative intercept on the length axis is the 'end effect' and the compliance is given by the slope. Stiffness is the inverse of compliance. Usually if the specimen is ten times longer than wide the end effect will be negligible. The big problem with tensile tests is clamping the specimen into the machine. Biological materials, going to relatively high strains, frequently become thin in the clamps. Pneumatic or other self-adjusting clamps are a great help. Use araldite or cyanoacrylate adhesive to glue coarse emery paper or aluminium tabs onto the end of the specimen. These provide extra friction and/or more support. For seaweed we found it necessary to wrap the ends of the specimens in absorbent tissue. If you have sufficient finances, use a freezing clamp for tissues which have plenty of water in them.

The measurement of tensile stiffness or strength is very simple. Mussel byssus threads are very easy to obtain and manipulate (Smeathers & Vincent 1979). Make cylinders of stiff Mylar about 3 cm diameter and 8 cm long, held together with paper clips or staples. Put one mussel in each tube then stack the tubes horizontally in a suitably irrigated aquarium. Within a day or so each mussel will have produced a large number of threads. Cut the threads where they leave the byssal root and remove the mussel to another Mylar tube where it can repeat the trick. The threads have corrugated and plain parts. What is the difference in stiffness, strength and extensibility of the two parts? Do they show differences in hysteresis behaviour? Or stress relaxation? Are they different when wet or dry? If you cut the Mylar from around the plaque which the mussel forms to stick the thread down you can test the attachment as well. How does its strength compare with that of the thread? Does it matter at what angle you pull it (hint: stick the Mylar onto a piece of bent aluminium so that you can clamp the aluminium in the machine and pull at the adhesive plaque at various angles)? How does the plaque break away? Is the force related to the area of the plaque? Compare mussels taken from sheltered and exposed areas of coast which are then allowed to make byssus threads as described above under controlled conditions. Are they the same? Use Mylars with different surface properties (measure by the contact angle of a drop of water). Do the plaques adhere equally well?

Hair and silk are also easy materials to test (Denny 1976, 1980). Spiders can be made to spin webs if you keep them in plastic lunch boxes. Line the lunch boxes with Mylar and other materials so that you can test the adhesion of the web to the different surfaces in much the same way as with the byssus plaques. Examine the adhesive joints in the SEM. Is there any difference with the differing kinds of surface? Does this correlate with strength of attachment? How does the spider know what to do? Modern shampoos are formulated for different types of hair and water. How do they affect the hair? Will it stretch more easily or be more easy to wave? Are there differences in different types of hair – e.g. hair from the head and from other parts of the body? Or hair of different colours? Or from different animals, breeds of sheep, etc. How effective are the various cross-linking agents used by hairdressers to induce 'permanent' waves?

The chalaza of a bird's egg presents an interesting problem in mechanical design. It is relatively easy to remove and is continuous with the membranes around the yolk and around the albumen, which makes it easy to clamp in the test machine. If the test is done fairly quickly the chalaza will not dry out. What is the modulus? What is the ultimate strain? Does the stress relax if it is held extended? What component is actually carrying the load? What can be the function of the chalaza?

There are quite a few plant fibres which are also easy to test. For instance, how does stiffness vary with the angle at which the cellulose is wound around the cell (measure the orientation using a polarising microscope)? Does tensile stiffness vary with lignification? Would you expect it to? And strength?

The wing of an adult insect has to unfold and stiffen when the insect undergoes its final ecdysis (Glaser & Vincent 1979). It is pumped up by fluid (no-one knows how – the wing will expand even if it is cut off) yet must not stretch when it's pulled out of the exuvia. How does the stiffness of the wing change during the time between the start of ecdysis and final expansion? Is it due to unfolding or stretching of the wing cuticle? Is the material elastic or plastic at various times during expansion? Is it equally stiff/extensible along as across the wing? What happens if you poison the wing tissue or keep it in an anoxic environment – e.g. nitrogen gas? What sort of mechanism might be controlling the extensibility?

8.2.2 Poisson's ratio – one-dimensional

There is a whole series of experiments which actually *need* to be done measuring Poisson's ratio on a variety of soft tissues. The technique is simple enough – use strips of material with marks on them (ink, specks of carbon, small pins stuck through) and stretch them by varying amounts, noting the way the material narrows as it is stretched. You will need to use a travelling microscope or cathetometer to measure strains. Plot Poisson's ratio against strain, using both engineering and real strain. Use skin from various parts of the body (from the back, the belly, cow teat, rat gut, etc). Change the length/width ratio of the specimens and see how this affects Poisson's ratio. Use knitted and woven materials (tights or stockings, linen) at various orientations as models. How do their Poisson's ratios vary with strain? How must the fibre be arranged to give these properties? One project with which we were involved was with a surgeon who removed tattoos. He was experimenting with different shapes of incision which would give better healing of the wound after the tattooed area of skin had been removed. He told the patients that he could remove only half the tattoo at a time and they would have to come back to have the second half removed. This always ensured that he could examine the progress of his initial operation. He stretched pieces of tattooed skin *in situ* and photographed them before and after stretching. The tattoo gave very nice fiducial points, so we were able to compare laboratory values for Poisson's ratios with those obtained *in vivo*.

   The intersegmental membrane of the mature female locust referred to in Section 4.5 has a very low Poisson's ratio, but there are probably other tubes which have to extend without narrowing and which therefore need to have odd Poisson's ratios and strange orientations of fibres. One obvious example is sea anemones. Gosline (1971) showed that the stiffness of *Metridium* mesogloea is greater circumferentially than longitudinally (Section 4.3), but he never measured Poisson's ratios properly, nor are there any values published. In this instance the orientation of fibres has been documented, but there is no indication of how the inner and outer layers of the mesogloea might interact. The siphon sheath of a number of molluscs which live in the sand or mud would be a good subject. We did briefly look at these and they seemed promising. Mount a section of the sheath in the universal test machine and measure transverse and longitudinal moduli, then Poisson's ratios. How do the ratios vary with extension? It is likely that the longitudinal one should fall at least as low as the locust membrane does. How is the sheath extended? What is the fibre phase which gives it its anisotropy? Does the Poisson's ratio vary from species to species according to extensibility? Are there any other bits of the mollusc which extend anisotropically? Do they all have strange orientations of fibres?

8.2.3 Fracture in tension

You can learn much about the design and ecology of a material using 'simple' fracture tests in which a notch is propagated across the material in tension (Jeronimidis 1976, Kitchener 1987, Vincent 1982, 1983, 1990). The important thing here is not just to plot notch-sensitivity curves but to watch the material while it fractures and to examine the fracture surfaces in the SEM. Look out for mechanisms which slow down the propagation of the crack tip or which deflect it. Ideally the notch should travel through the material relatively slowly so that there is time for all the elastic energy within the material to find its way to the fracture surface. Reduce the speed of the crosshead on the test machine accordingly. Plant material is very easy to work with. It makes sense to experiment with materials which are protective in some way. Despite the large amount of effort expended by horticulture researchers, not one has yet studied the fracture toughness of fruit skins. Strength and stiffness, yes. But these tell only part of the protective properties of a skin. Compare tomato and apple skins. Tomato skin breaks very easily once a crack has started, but it looks as if apple skin retains one or two layers of cells inside the outer layer and this retards the propagation of cracks. Look at the skins of a variety of fruits. Do they fall into the category of tomato or apple skins? Calculate the works of fracture. Do different varieties of apple have skins fracturing in a different way? Or with different works of fracture? How far is this related to the way the parenchyma cells stick to the skin? Compare fracture of skin from a variety of apple with differing amounts of apple flesh left attached.

Another membrane which gives very nice results is the egg-case of various types of shark and dogfish. The case is made of orthogonally woven collagen fibres. You could also try the inner membrane of an egg or, if you know a friendly gynaecological surgeon, amniotic membrane.

Getting into more specialized areas of fracture, using the double cantilever beam (Atkins & Mai 1985, Bertram & Gosline 1986) it is quite fun to measure the fracture properties of various nuts. We originally tested the fracture properties of the coconut (Vincent 1990), wondering whether the coconut would make a good crash helmet for skateboarders! We later compared our figures with some for the much-vaunted macadamia nut (Jennings & Macmillan 1984), finding little difference between the two. So how do the mechanical properties of nut shells vary? It is not difficult to machine small pieces for a variety of tests other than the double cantilever beam. Jennings and Macmillan cut circular sections transversely across the nut, removing the endosperm and treating the

the specimen like a ring (Roark & Young, 1975). How do the stiffness and toughness of the nut vary with lignification, morphology of the fracture surface or the abilities or strength of animals likely to eat it?

8.3 COMPRESSION TESTS

8.3.1 Compressive stiffness and strength

The measurement of stiffness in compression is not quite so simple as at first it appears. The force must be fed properly into the specimen. It is more or less impossible to prepare a stiff specimen with parallel ends which can be compressed evenly between two plates. It is necessary to feed the force in by shear, which entails embedding the ends of the specimen in end blocks which are supported on balls (Fig. 8.1). This ensures that force is fed in evenly and that the specimen orientates itself accurately and automatically. The length/width ratio must be less than 10, preferably nearer 5. If the test piece is too long the specimen will bend and buckle, deforming in Euler (pronounced Oy-ler) mode (see below). Compression stiffness probably needs to be measured in this way only with ceramic materials. How do compressive and tensile stiffness and strength of (say) bone compare? With woody and other cellular materials the initial part of compression flattens out any unevenness at the ends of the test-piece and the test proceeds more or less controllably. Take a branch of a tree which has been growing sideways, horizontally. Compare the behaviour of the wood on the top and the bottom of the branch when tested in compression. How do the

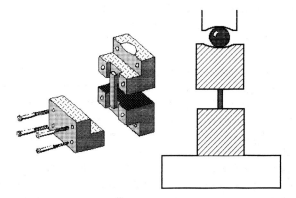

Figure 8.1 Rig for compression test of small stiff specimens. Left: the specimen is laid in slots in the end pieces and embedded in araldite (not shown). The end pieces are completed with an extra piece screwed in position only one shown) and the araldite is allowed to cure. Right: the specimen is loaded via a steel ball which sits in dimples. The steel ball makes it easier to get the specimen properly in line with the applied force.

stiffness and failure characteristics compare (refer to Gibson & Ashby 1988)? Remember that the wood on the upper side of the branch grew, and expected always to be, in tension. Failure in compression of a cellular material will most commonly start by local buckling. Compare failure in wood (an open cellular material) with that in a closed cellular material such as rhubarb petiole (chosen because sections can be cut transversely along it which will be more or less uniform). How do stiffness and strength vary with changes in turgor? Does the mode of failure change as turgor changes? Does the mode of failure change depending whether or not the epidermis is left on? What is the function of the epidermis (see Section 8.4 on beam tests).

8.3.2 Euler buckling

If the compression test piece is more than about 30 times longer than wide (a strut rather than a column) it will tend to deflect sideways under the end load and buckle away to one side (Fig. 8.2). This is Euler buckling (Timoshenko & Goodier 1970). The load at which buckling occurs depends in part on the end constraints. It is easiest if the ends are able to seat in dimples so that they automatically align themselves within the machine. A number of biological structures are designed as struts – porcupine quills, long ovipositors of parasitic

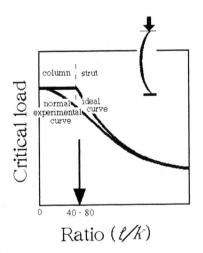

Ratio (ℓ/k)

Figure 8.2 Euler buckling. Depending on the material, the transition between compressive column and strut comes at a length/radius of gyration ratio of 40 to 80 ('ideal' curve). In real life the transition is never so sharp since it is impossible to align the strut/column absolutely accurately ('normal' curve). Refer to Timoshenko & Goodier (1970) for a fuller explanation of this mode of deformation and the associated method of Southwell for analysing experimental results.

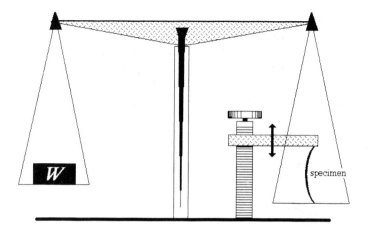

Figure 8.3 A chemical balance modified for use as a test machine. The specimen is being loaded in compression and is deforming elastically in Euler buckling. The degree of deformation is controlled by an adjustable arm which forms the upper end piece (right). Both upper and lower end pieces are dimpled to locate the ends of the specimen. The force is balanced by weights in the left pan. This sort of machine can be modified in a number of ways for a variety of experiments.

wasps, long plant stems and possibly the legs of Tipulid flies and phalangid arachnids. There are a number of subtleties in Euler buckling but it is a very useful form of test and can have great biological relevance (Vincent & Owers 1986). If the lateral deflection of the strut is measured it is possible to plot the force deflection data as a straight line (Southwell plot – see Timoshenko & Gere 1961) which enables the critical Euler load to be calculated easily. In turn this allows the stiffness of the material to be calculated. Test the various spines from a number of different species of porcupine. Are they all the same mechanically? If they have different types of material down the middle, does this affect either stiffness or failure load? Does failure occur on the compressive or tensile surface? Calculate the compressive failure stress (it might be easier to do this in 4-point bending – see Vincent & Owers 1986 & Fig. 8.4). Does the quill or spine fail as a structure (i.e. by local buckling) or as a material (i.e. by reaching its compressive strength)? Observe a parasitic wasp which has a very long ovipositor (e.g. *Megarhyssa*). Does it run into problems of Euler buckling? Similarly, are the long legs which many sorts of arthropod have in any danger of Euler buckling? You may need to test the legs and ovipositor in a rather more delicate machine than the average universal test machine, although these can typically measure to a fraction of a gram. It is not too difficult to modify a chemical balance as a test machine, using a movable platform mounted on a rack-and-pinion to change the position of the specimen, bringing the balance back to null by the addition of weights or winding and unwinding a chain from the end of the balance beam (Fig. 8.3).

(a)

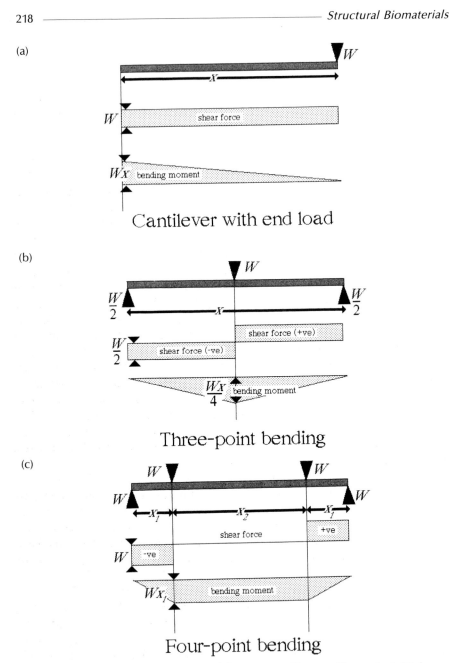

Cantilever with end load

(b)

Three-point bending

(c)

Four-point bending

Figure 8.4 Three different types of beam test. The cantilever test (a) is very convenient in many situations; the three-point bending test (b) is also useful, especially for controlled fracture tests in which the specimen is notched in the middle. The four-point bending test (c) is especially useful for measuring ultimate compressive properties since the bending moment is constant over the middle span and there are no stress concentrations. Refer to Roark & Young (1975).

8.4 BEAM TESTS

There are several ways in which a beam can be deformed (Fig. 8.4; Stephens 1970; Roark & Young 1975). Three- and four-point bending tests, which can be performed in the universal testing machine, are very convenient since there is little problem with mounting the specimen. However, the specimen has to have a sufficiently hard surface not to deform and you have to be sure whether you are measuring shear or Young's modulus (Roark & Young 1975, Jackson *et al.* 1988). The safest thing is to perform a number of tests at differing span-to-depth (S/D) ratios and see at what S/D ratio the Young's modulus plateaus out. It is very salutary to measure the stiffness of internodal sections of the stem of various sized hogweeds (umbellifers of the genus *Heracleum* and others closely related) and to compare their stiffness and degree of lignification. Does the lignification make any difference? What's it doing then?? Or is it taking over from turgor in some way? Another nice system is the petiole of a rhubarb leaf. This is a nice large beam which is more or less uniform along its length. Measure its stiffness in bending, then remove the epidermis from the upper side. Measure stiffness in bending again, first with the epidermis on the tensile side, then on the compression side. What is the difference? Measure the stiffness of the epidermis in a tensile test. Use composite beam theory to calculate the stiffness of the petiole with a stiff membrane all around it. Compare these results with the compression tests on sections of whole petiole. The petiole is a skin-core structure. Do some similar tests with leaves of reeds and irises, which are also skin-core structures (Gibson & Ashby 1988, Gibson *et al.* 1988). How does the mechanical design of the leaf vary with the ecology of the plant? For instance, compare the deflections and failure loads of leaves with the sorts of deflections and rate of failure due to the wind blowing. Hence get some idea of safety factors (Alexander 1981, 1982). We found that sweet rush has much softer and less 'safe' leaves than the iris, but they tend to grow with much more support, both from the surrounding water and because they tend to grow in denser clumps. You can do the same sort of thing with flowering stems of the dandelion, *Taraxacum officinale*. Compare stems grown in shady woodland with those grown in an open field or meadow. How do the mechanical properties change when the seed head develops and the stem grows even higher?

A number of trees 'weep'; the most familiar is the weeping willow. Why? Compare the stiffness of twigs and branches of a number of species of weeping and non-weeping willows. Does the willow weep because its wood is different? Or is there less of it? Or are the branches and twigs differently shaped? How does this compare with weeping ash and cherry? And with fastigiate trees?

8.5 SUNDRY FRACTURE TESTS

There are ways in which a material can be fractured in tension other than by holding on to the two ends and pulling. One of the most useful involves pushing a wedge into the material. As the wedge penetrates the material it initially cuts it, but (depending on the included angle) later the sides of the wedge push against the sides of the crack storing elastic energy in the bent sides (Fig. 8.5). This energy can now be fed into the advancing crack which then moves ahead of the tip of the wedge in the same way as it would in a tensile test. The wedge test is very useful for particles down to 2 mm across (use a hard-backed razor blade to provide the wedge) and for test pieces which would otherwise be difficult to stretch (apple parenchyma and other cellular materials). The work of fracture is easily measured: the area under the force-deflection graph is the work; this is expressed in terms of the area cleaved (Atkins & Mai 1985). How does the work of fracture vary with the variety of apple and the orientation of the crack (Khan 1988; Vincent 1989, 1990)? Does this correlate at all with 'texture' ('crispness', 'toughness') which you perceive when you eat the apple?

FORCE

Figure 8.5 The wedge test (see text).

A related fracture test involves cutting. Although this was originally studied using an instrumented microtome (Atkins & Vincent 1984; Willis 1989) it is quite possible to mount a sharp razor in a universal testing machine. The work is then a product of the force and the distance, as before. But it is also possible to pick up variation of force during the cut which can be related to variations in the morphology of the test piece. So an instrumented knife can act as a sort of mechanical microscope! Compare the work of fracture measured in cutting with that measured with a free-running crack. Why is there any difference? Change the turgor of plant parenchyma (potato is ideal) using solutions of mannitol (0% to 15%) and measure the work of fracture by cutting and with a tensile notch-sensitivity test (or the wedge test, though this is a bit more complex with a less turgid specimen). Do the two methods give the same answer? If not, why not? You can make vegetables more crisp by chilling them. Try this with carrot. Can you detect this with the cutting test? Cut wood in this manner (you need to have a very stiff and well-supported blade to do this). Is there any difference between tension and compression woods?

We tried measuring the ease with which squirrels could strip bark from trees. The branch is held down to the base of the universal test machine and a piece of bark peeled off and attached to a long wire from the load cell (a long wire

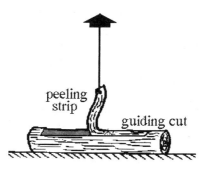

Figure 8.6 Measuring the work required to peel bark.

is used so that the angle of pulling is not changed very much as the bark peels away). A strip is then defined by a couple of razor cuts 1 cm or so apart, and the bark is then peeled off (Fig. 8.6; Vincent 1990). Once again, work is measured as force × distance and normalized to the area cleaved. How does the fracture energy vary with different species of tree and at different times of year? How does this relate to the incidence of damage by squirrels and other pests? Which layer fractures? Might this be seasonal? Why?

8.6 SUNDRY OTHER EXPERIMENTS

There are numerous other experiments which are very difficult to categorize. One which yielded some nice results concerned the viscoelastic properties of the hinge ligament of molluscs. The argument runs like this. You would expect the ligament in an actively swimming mollusc such as *Pecten* to have only a small or non-existent viscous component. But the ligament of a sessile mollusc has much more hysteresis (Kahler *et al.* 1976). So it should show much greater stress relaxation. Using the apparatus of Fig. 8.7 it is possible to measure this. Our initial results looked promising. Quantify the degree of stress relaxation in the hinge of a number of molluscs. Keep the temperature at a suitable value for that particular mollusc. Does the rate of relaxation relate to the activity of the mollusc? Remember that a sessile mollusc needs to open and close its valves whether it is sitting on a rock or in the sand.

Although hardness is measured with an indentation test, this may not always be related to wear which is relatively ill-defined. A micro-wear test is very easy to perform using a disc of fine corundum paper glued to a turntable driven by a small motor. The object to be tested is held against the rotating corundum disc, at a constant distance from the centre, under a known preload. The disc should be irrigated with water to wash away the general debris and keep the cutting surface clear. It takes some skill to measure wear, especially on such a small object as a tooth from the radula of a mollusc such as *Patella*. But it

Force
transducer

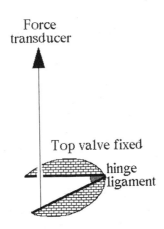

Top valve fixed

hinge
ligament

Figure 8.7   One way of measuring stress-relaxation of the mollusc hinge ligament in compression. The top valve is fixed (e.g. by long bolts to the base of the test machine) and the lower valve is connected to the cross-head of the test machine by a long piece of piano wire (0.04 mm diameter – a banjo 1st or 5th string – is ideal).

is possible. Measure the wear on teeth taken from different positions along the radula. How does it vary with different species of mollusc? Whilst *Patella* has silica and goethite in its teeth, *Chiton* has the harder magnetite. Compare radular teeth from molluscs feeding from different substrates (molluscs living on a rocky shore or in the garden). The beak of the octopus and related molluscs is also rather hard at the tip. How does this compare in terms of wear? We looked for metals in an octopus beak but could not find any. But it is worth looking again if you find it to be really hard.

8.7   POISSON'S RATIO IN TWO DIMENSIONS

The act of cutting a sample from a sheet of tissue removes edge constraints which may be important. It is fairly easy to find out by measuring the Poisson's ratio of a tube and a sheet cut from the tube. The most convenient tube is a length of gut from a rat or similar animal, though we have used cow teats (their properties were of interest to a designer of milking machines; tests on their Poisson's ratio involved inflating them, which evoked some ribald comments, especially when it was observed that, in order to do the job properly, we were putting a condom *inside* the teat!). The gut should be tied off at one end and the other attached to a variable head of water, so that the pressure inside the tube can be measured directly in terms of cm of water. Vary the pressure and measure changes in length and diameter of the tube. These sorts of measurement can be applied to other tubes such as arteries (Dobrin & Doyle 1970, Gosline & Shadwick 1982). What are the implications of these results for guts and arteries?

8.8 COMPUTER MODELLING

Obviously it makes sense either to use a computer spreadsheet or write a program for the analysis of results from experiments on biological materials. But a computer program can also be of help to the biologist (who is frequently more at home with a graph than with a mathematical formula) to use some of the basic formulae (as in Chapter 1) as the basis of programs to illustrate (for example) the effect of adding together several spring-and-dashpot elements, obtaining a variety of relaxation spectrums. With the awful variability and complex shapes of biological materials, it can also be very useful to program a spreadsheet using whatever the simple theoretical model might be, then doing 'what-if?' calculations to discover where the inaccuracies and perturbations lie. In this way one can arrive at a better feeling for the analysis used and see where it is necessary to make more accurate measurements and where assumptions might be crucial.

Bibliography

Aizawa M, Suzuki S, Suzuki T & Toyoma H (1973) The properties of water in macromolecular gels. VI. The relationship between the rheological properties and the states of water in macromolecular gels. *Bull. Chem. Soc. Japan* **46**, 1638–1640.

Alexander RMcN (1962) Viscoelastic properties of the body-wall of sea anemones. *J. Exp. Biol.* **39**, 373–386.

Alexander RMcN (1966) Rubber-like properties of the inner hinge-ligament of Pectinidae. *J. Exp. Biol.* **44**, 119–130.

Alexander RMcN (1981) Factors of safety in the structure of animals. *Sci. Prog. Oxf.* **67**, 109–130.

Alexander RMcN (1982) *Optima for Animals.* Edward Arnold, London.

Alexander, RMcN, Brandwood A, Currey JD & Jayes AS (1984) Symmetry and precision of control of strength in limb bones of birds. *J. Zool., Lond.* **203**, 135–143.

Alexander, RMcN, Bennet MB & Ker RF (1986) Mechanical properties and function of the paw pads of some mammals. *J. Zool., Lond.* **209**, 405–419.

Andersen SO (1971) Resilin. In *Comparative Biochemistry* (ed. M Florkin & EH Stotz), 26C, pp. 633–657. Elsevier, Amsterdam.

Andersen SO (1986) Sclerotisation and tanning of the cuticle. In *Comparative Insect Physiology, Biochemistry and Pharmacology* (ed. GA Kerkut & LI Gilbert), vol. 3, pp. 59–74. Pergamon, Oxford.

Andersen SO & Weis-Fogh T (1964) Resilin, a rubber-like protein in arthropod cuticle. *Adv. Insect Physiol.* **2**, 1–65.

Arai KM, Takahashi R, Yokote Y & Akahane K (1983) Amino acid sequence of feather keratin from fowl. *Eur. J. Biochem.* **132**, 501–507.

Aspden RM (1986) Relation between structure and mechanical behaviour of fibre-reinforced composite materials at large strains. *Proc. R. Soc. Lond.* A **406**, 287–298.

Aspden RM (1988) The theory of fibre-reinforced composite materials applied to changes in the mechanical properties of the cervix during pregnancy. *J. Theor. Biol.* **130**, 213–221.

Astbury WT & Woods HJ (1933) X-ray data on the structure of hair, wool and related fibres. II The molecular structure and elastic properties of hair keratin. *Phil. Trans. R. Soc. Lond.* A **232**, 333–394.

Aszenzi A & Bell GH (1972) Bone as a mechanical engineering problem. In *The Biochemistry and Physiology of Bone* (ed. GH Bourne), pp. 311–352. Academic Press, London, New York.

Atkins AG & Mai Y-M (1985) *Elastic and Plastic Fracture: Metals, Polymers, Ceramics, Composites, Biological Materials.* Ellis Horwood, Chichester.

Atkins AG & Vincent JFV (1984) An instrumented microtome for improved histological sections and the measurement of fracture toughness. *J. Mat. Sci. Lett.*, **3**, 310–312.

Bacon GE, Bacon PJ & Griffiths RK (1979) Stress distribution in the scapula studied by neutron diffraction. *Proc. R. Soc. Lond.* B **204**, 355–362.

Bacon GE, Bacon PJ & Griffiths RK (1981) The unseen stresses on the human frame. *New Scientist* 17 December, 796–799.

Barrett TW (1984) The materials science opportunities in biotechnology. *Proceedings of Biotech '84 USA*, Online Publications, Pinner, UK, pp 103–126.

Behiri JC & Bonfield W (1982) Fracture mechanics of cortical bone. In *Biomechanics: Principles and Applications* (ed. R Huiskes, D Van Campen & J De Wijn), pp. 247–251. Martinus Nijhoff, Den Haag.

Benedict CV (1987) Method for making dopa-containing bioadhesive proteins from tyrosine-containing proteins. European Patent Application 87104853.4.

Bennet MB & Alexander, RMcN (1987) Properties and function of extensible ligaments in the necks of turkeys (*Meleagris gallopavo*) and other birds. *J. Zool., Lond.* **212**, 275–281.

Bennet MB & Stafford JA (1988) Tensile properties of calcified and uncalcified avian tendons. *J. Zool., Lond.* **214**, 343–351.

Bennet MB, Ker RF & Alexander, RMcN (1987) Elastic properties of structures in the tails of cetaceans (*Phocaena* and *Lagenorhynchus*) and their effect on the energy cost of swimming. *J. Zool., Lond.* **211**, 177–192.

Bennet MB, Ker RF, Dimery NJ & Alexander, RMcN (1986) Mechanical properties of various mammalian tendons. *J. Zool., Lond.* **209**, 537–548.

Bennet-Clark HC (1975) The energetics of the jump of the locust, *Schistocerca gregaria*. *J. Exp. Biol.* **63**, 53–83.

Bennet-Clark HC & Lucey ECA (1967) The jump of the flea. *J. Exp. Biol.* **47**, 59–76.

Bereiter-Hahn J, Matoltsy AG & Richards KS (1984) *Biology of the Integument*. Springer-Verlag, Berlin.

Berman A, Addadi L & Weiner S (1988) Interactions of sea-urchin skeleton macromolecules with growing calcite crystals – a study of intracrystalline proteins. *Nature, Lond.* **331**, 546–548.

Berry JP & Morrell SH (1974) Liquid rubbers and the problems involved in their application. *Polymer* **15**, 521–526.

Bertram JEA & Gosline JM (1986) Fracture toughness design in horse hoof keratin. *J. Exp. Biol.* **125**, 29–47.

Biewener A, Alexander, RMcN & Heglund NC (1981) Elastic energy storage in the hopping of kangaroo rats (*Dipodomys spectabilis*). *J. Zool., Lond.* **195**, 369–383.

Birchall JD (1989) The importance of the study of biominerals to materials technology. In *Biomineralization, Chemical and Biochemical Perspectives* (ed. S Mann, J Webb & RJP Williams). VCH Publishers, New York.

Blackwell J & Gelman RA (1975) Polypeptide-mucopolysaccharide interactions. In *Fibrous Biopolymers* (ed. EDT Atkins & A Keller), pp. 93–108. Butterworth, London.

Blackwell J & Weih MA (1980) Structure of chitin protein complexes: ovipositor of the ichneumon fly. *Megarhyssa. J. Mol. Biol.* **137**, 49–60.

Blyski DI, Kriewall TJ, Akkas N & Melvin JW (1986) Mechanical behaviour of fetal dura mater under large deformation biaxial tension. *J. Biomechanics* **19**, 19–26.

Bond GM, Scott VD & Board RG (1986) Correlation of mechanical properties of avian eggshells with hatching strategies. *J. Zool., Lond.* **209**, 225–237.

Bonfield W (1987) Materials for the replacement of osteoarthritic hip joints. *Met. Mat.* **3**, 712–716.

Bonfield W & Datta PK (1976) Fracture toughness of compact bone. *J. Biomechanics* **9**, 131–134.

Bonfield W & Grynpas MD (1977) Anisotropy of the Young's modulus of bone. *Nature, Lond.* **270**, 453–454.

Bonfield W, Grynpas MD & Young RJ (1978) Crack velocity and the fracture of bone. *J. Biomechanics* **11**, 473–479.

Bouligand Y & Giraud-Guille MM (1985) Spatial organisation of collagen fibrils in skeletal tissues: analogies with liquid crystals. In *Biology of Invertebrate and Lower Vertebrate Collagens* (ed. A Bairati & R Garrone). Plenum, Paris.

Boyde A (1972) Scanning electron microscope studies of bone. In *The Biochemistry and Physiology of Bone* (ed. GH Bourne), pp. 259–310. Academic Press, London, New York.

Braem M, van Doren VE, Lambrechts P & Vanherle G (1987) Determination of Young's modulus of dental composites: a phenomenological model. *J. Mat. Sci.* **22**, 2037–2042.

Brear K & Currey JD (1976) Structure of a sea urchin tooth. J. Mat. Sci. **11**, 1977–1978.

Bretz PE, Hertzberg RW & Manson JA (1979) Fatigue crack propagation in crystalline polymers: effect of moisture in nylon 66. *J. Mat. Sci.* **14**, 2482–2492.

Broom ND (1984a) Further insights into the structural principles governing the function of articular cartilage. *J. Anat.* **139**, 275–294.

Broom ND (1984b) The altered biomechanical state of human femoral head osteoarthritic articular cartilage. *Arthr. Rheum.* **27**, 1028–1039.

Brown CH (1975) *Structural Materials in Animals*. Pitman, London.

Brunet PCJ & Coles BC (1974) Tanned silks. *Proc. R. Soc.* B **187**, 133–170.

Brush AH (1983) Self-assembly of avian psi-keratins. *J. Prot. Chem.* **2**, 63–75.

Bubner E, Meyer DE, Schillak L & Schumacher H (1988) Bauprozesse im Meerwasser auf elektrochemische-biogener Grundlage. *Natuerliche Konstructionen (Proc. Sonderforschungsbereiches 230, Stuttgart)*, **1**, 95–105.

Burkhardt A, Hansmann W, Maerkel K & Niemann H-J (1983) Mechanical design in spines of Diadematoid echinoids (Echinodermata, Echinoidea). *Zoomorph.* **102**, 189–203.

Calvert PD (1989) Bio-mimetic processing of ceramics and composites. In *Proceedings of International Workshop on Intelligent Materials*, pp. 63–72. The Society of Non-Traditional Technology, Tokyo.

Calvert PD & Mann S (1988) Review: Synthetic and biological composites formed by *in situ* precipitation. *J. Mat. Sci.* **23**, 3801–3815.

Cook J & Gordon JE (1964) A mechanism for the control of crack propagation in all-brittle systems. *Proc. R. Soc. Lond.* A **282**, 508–520.

Cowan PM, North ACT & Randall JT (1958) X-ray diffraction studies of collagen fibres. In *Fibrous Proteins and their Biological Significance* (*Symp. Soc. Exp. Biol. 9*), (ed. R Brown & JF Danielli), pp. 115–126.

Currey JD (1975) The effects of strain rate, reconstruction and mineral content on some mechanical properties of bovine bone. *J. Biomechan.* **8**, 81–86.

Currey JD (1977) Mechanical properties of mother of pearl in tension. *Proc. R. Soc. Lond.* B, **196**, 443–463.

Currey JD (1979) Mechanical properties of bone tissues with greatly differing functions. *J. Biomechanics* **12**, 313–319.

Currey JD (1980) Mechanical properties of mollusc shell. In *The Mechanical Properties of Biological Materials* (*Symp. Soc. Exp. Biol.* 34) (ed. JFV Vincent & JD Currey), pp. 75–97. Cambridge University Press, Cambridge.

Currey JD (1984a) Effects of differences in mineralisation on the mechanical properties of bone. *Phil. Trans. R. Soc. Lond.* B **304**, 509–518.

Currey JD (1984b) *The Mechanical Adaptations of Bones*. Princeton University Press, Princeton.

Currey JD & Brear K (1984) Fatigue fracture of mother-of-pearl and its significance for predatory techniques. *J. Zool., Lond.* **203**, 541–548.

Currey JD & Kohn AJ (1976) Fracture in the crossed-lamellar structure of *Conus* shells. *J. Mat. Sci.* **11**, 1616–1623.

Currey JD & Taylor JD (1974) The mechanical behaviour of some molluscan hard tissues. *J. Zool. Lond.*, **173**, 395–406.

Currey JD, Nash A & Bonfield W (1982) Calcified cuticle in the stomatopod smashing limb. *J. Mat. Sci.* **17**, 1939–1944.

Currey JD & Pond, Caroline M (1989) Mechanical properties of very young bone in the axis deer (*Axis axis*) and humans. *J. Zool., Lond.* **218**, 59–67.

Daculsi G, Menanteau J, Kerebel LM & Mitre D (1984) Length and shape of enamel crystals. *Calcif. Tiss. Int.* **36**, 550–555.

Danilatos G & Feughelman M (1979) Dynamic mechanical properties of alpha-keratin fibres during extension. *J. Macromol. Sci. – Phys.* B **16**, 581–602.

Denny MW (1976) The physical properties of spiders' silk and their role in the design of orb-webs. *J. Exp. Biol.* **65**, 483–506.

Denny MW (1980) Silks – their properties and functions. In *The Mechanical Properties of Biological Materials* (*Symp. Soc. Exp. Biol.* 34) (ed. JFV Vincent & JD Currey), pp. 247–272. Cambridge University Press, Cambridge.

Denny MW (1984) Mechanical properties of pedal mucus and their consequences for gastropod structure and performance. *Am. Zool.* **24**, 23–36.

Dickerson RE & Geis I (1969). *The Structure and Action of Proteins.* WA Benjamin, Menlo Park, California.

Dimery NJ & Alexander, RMcN (1985) Elastic properties of the hind foot of the donkey, *Equus asinus. J. Zool., Lond.* **207**, 9–20.

Dimery NJ, Alexander, RMcN & Deyst KA (1985) Mechanics of the ligamentum nuchae of some artiodactyls. *J. Zool., Lond.* **206**, 341–351.

Dobrin PB & Doyle JM (1970) Vascular smooth muscle and the anisotropy of dog carotid artery. *Circ. Res.* **27**, 105–119.

Dorrington KL (1980) The theory of viscoelasticity in biomaterials. In *The Mechanical Properties of Biological Materials* (*Symp. Soc. Exp. Biol.* 34) (ed. JFV Vincent & JD Currey), pp. 289–314. Cambridge University Press, Cambridge.

Druhala M & Feughelman M (1974) Dynamic mechanical loss in keratin at low temperatures. *Colloid Polym. Sci.* **252**, 381–391.

Easterling KE, Harrysson R, Gibson LJ & Ashby MF (1982) On the mechanics of balsa and other woods. *Proc. R. Soc. Lond.* A **383**, 31–41.

Eastoe JE (1971) Dental enamel. In *Comprehensive Biochemistry* (ed. M Florkin & EH Stotz), 26C, pp. 785–834. Elsevier, Amsterdam.

Emlet RB (1982) Echinoderm calcite: a mechanical analysis from larval spicules. *Biol. Bull.* **163**, 264–275.

Espey LL (1967) Tenacity of porcine Graafian follicle as it approaches ovulation. *Am. J. Physiol.* **212**, 1397–1401.

Evans AG & Charles EA (1976) Fracture toughness determinations by indentation. *J. Am. Ceram. Soc.* **59**, 371–372.

Eylers JP & Greenberg AR (1989) Swelling behaviour of the catch connective tissue in holothurian body wall. *J. Exp. Biol.* **143**, 71–85.

Falk S, Hertz CH & Virgin HI (1958) On the relation between turgor pressure and tissue rigidity. I. *Physiol. Plant.* **11**, 802–817.

Farber JN & Farris RJ (1987) Model for prediction of the elastic response of reinforced materials over wide ranges of concentration. *J. Appl. Polym. Sci.* **34**, 2093–2104.

Ferry JD (1970) *Viscoelastic Properties of Polymers.* John Wiley, New York.

Feughelman M (1979) A note on the role of the microfibrils in the mechanical properties of alpha-keratins. *J. Macromol. Sci. – Phys.* B **16**, 155–162.

Fox PG (1980) The toughness of tooth enamel, a natural fibrous composite. *J. Mat. Sci.* **15**, 3113–3121.

Fraenkel G & Rudall KM (1940) A study of the physical and chemical properties of the insect cuticle. *Proc. R. Soc. Lond.* B **129**, 1–35.

Franzblau C (1971) Elastin. In *Comparative Biochemistry* (ed. M Florkin & EH Stotz), 26C, pp. 659–712. Elsevier, Amsterdam.

Fraser RDB & MacRae TP (1980) Molecular structure and mechanical properties of keratins. In *The Mechanical Properties of Biological Materials* (*Symp. Soc. Exp. Biol.* 34) (ed JFV Vincent & JD Currey), pp. 211–246. Cambridge University Press, Cambridge.

Fraser RDB, MacRae TP, Parry DAD & Suzuki E (1971) The structure of feather keratin. *Polymer* 12, 35–56.

Fraser RDB, MacRae TP, Rowlands RJ & Tulloch PA (1976) Molecular structure of keratin. In *Proc. 5th Int. Wool Textile Res. Conf.* (ed. K Ziegler), vol. 2, pp. 80–88. Deutches Wollforschungsinstitut, Aachen.

Gibbs DA, Merrill EW & Smith KA (1968) Rheology of hyaluronic acid. *Biopolymers* 6, 777–791.

Gibson LJ (1984) *The Mechanical Behaviour of Cancellous Bone*, pp. 1–17 plus figures. MIT Dept. of Civil Engineering, Cambridge, Mass. Publication No. R84–15.

Gibson LJ & Ashby MF (1988) *Cellular Solids, Structure and Properties*. Pergamon, Oxford.

Gibson T, Stark H & Kenedi RM (1971) The significance of Langer's lines. *Trans. 5th Int. Congr. Plastic & Reconstruct. Surgery* (ed. JT Hueston). Butterworth, Australia.

Gibson LJ, Easterling KE & Ashby MF (1981) The structure and mechanics of cork. *Proc. R. Soc. Lond.* A 377, 99–117.

Gibson LJ, Ashby MF & Easterling KE (1988) Structure and mechanics of the iris leaf. *J. Mat. Sci.* 23, 3041–3048.

Giraud-Guille MM (1988) Twisted plywood architecture of collagen fibrils in human compact bone osteons. *Calcif. Tissue Int.* 42, 167–180.

Glaser AE & Vincent JFV (1979) The autonomous inflation of insect wings. *J. Insect Physiol.* 25, 315–318.

Glimcher MJ (1984) Recent studies of the mineral phase in bone and its possible linkage to the organic matrix by protein-bound phosphate bonds. *Phil. Trans. R. Soc. Lond.* B 34, 479–508.

Gordon JE (1976) *The New Science of Strong Materials, or Why You Don't Fall Through the Floor*, 2nd edn. Penguin, Harmondsworth.

Gordon JE (1978) *Structures, or Why Things Don't Fall Down*. Penguin, Harmondsworth.

Gordon JE (1980) Biomechanics: the last stronghold of vitalism. In *The Mechanical Properties of Biological Materials* (*Symp. Soc. Exp. Biol.* 34) (ed. JFV Vincent & JD Currey), pp 1–11. Cambridge University Press, Cambridge.

Gordon JE (1987). *The Science of Structures and Materials*. Freeman, New York.

Gordon JE & Jeronimidis G (1980) Composites with high work of fracture. *Phil. Trans. R. Soc. Lond.* A 294, 545–550.

Gosline JM (1971) Connective tissue mechanics of *Metridium senile* II Viscoelastic properties and a macromolecular model. *J. Exp. Biol.* 55, 775–795.

Gosline JM (1980) The elastic properties of rubber-like proteins and highly extensible tissues. In *The Mechanical Properties of Biological Materials* (*Symp. Soc. Exp. Biol.* 34) (ed. JFV Vincent & JD Currey), pp. 331–357. Cambridge University Press, Cambridge.

Gosline JM & French CJ (1979) Dynamic mechanical properties of elastin. *Biopolymers* 17, 677–695.

Gosline JM & Shadwick RE (1982) The biomechanics of the arteries of *Nautilus*, *Nototodarus* and *Sepia*. *Pacific Science* 36, 283–296.

Gosline JM, Denny MW & DeMont ME (1984) Spider silk as rubber. *Nature, Lond.* 309, 551–552.

Gosline JM, DeMont ME & Denny MW (1986) The structure and properties of spider silk. *Endeavour* **10**(1), 37–43.

Gregg K & Rogers GE (1984) Feather keratin: composition, structure and biogenesis. In *Biology of the Integument* (ed. J Bereiter-Hahn, AG Matoltsy & KS Richards), pp. 666–694. Springer-Verlag, Berlin.

Gregg K, Wilton SD, Parry DAD & Rogers GE (1984) A comparison of genomic coding sequences for feather and scale keratins: structural and evolutionary implications. *EMBO J.* **3**, 175–178.

Griffith AA (1921) The phenomena of rupture and flow in solids. *Phil. Trans. R. Soc.* A **221**, 163–198.

Guild FJ, Harris B & Atkins AG (1978) Cracking in layered composites. *J. Mat. Sci.* **13**, 2295–2299.

Hackman RH & Goldberg M (1979) Some conformational studies of *Calliphora vicina* larval cuticular protein. *Insect Biochem.* **9**, 557–561.

Harris B (1980) The mechanical behaviour of composite materials. In *The Mechanical Properties of Biological Materials* (*Symp. Soc. Exp. Biol.* 34) (ed. JFV Vincent & JD Currey), pp. 37–74. Cambridge University Press.

Harwood JAC, Mullins L & Payne AR (1965) Stress softening in natural rubber vulcanisates. Part 2. Stress softening effects in pure gum and fill-loaded rubbers *J. Appl. Polym. Sci.* **9**, 3011–3021.

Hashin Z & Rosen BW (1964) The elastic moduli of fiber-reinforced materials. *J. Appl. Mech.* **31**, 223–232.

Hearle JWS, Chapman BM & Senior GS (1971) The interpretation of the mechanical properties of wool. *Applied Polymer Symposium* **18**, 775–794.

Hearmon RFS (1946) The elastic constants of anisotropic materials. *Rev. Mod. Phys.* **18**, 409–440.

Hepburn HR & Chandler HD (1976) Material properties of arthropod cuticles: the arthrodial membranes. *J. Comp. Physiol.* **109**, 177–198.

Hickey DK & Hukins DWL (1980) X-ray diffraction studies of the arrangement of collagenous fibres in human fetal intervertebral disc. *J. Anat.* **131**, 81–90.

Hilbertz W (1988) Growing and fading structures: experiments, applications, ideas. *Natuerliche Konstructionen* (*Proc. Sonderforschungsbereiches 230, Stuttgart*), **1**, 107–114.

Hillerton JE (1980) The hardness of locust incisors. In *The Mechanical Properties of Biological Materials* (*Symp. Soc. Exp. Biol.* 34) (ed. JFV Vincent & JD Currey) pp. 483–484, Cambridge University Press, Cambridge.

Hillerton JE & Vincent JFV (1979) The stabilisation of insect cuticles. *J. Insect Physiol.* **25**, 957–963.

Hillerton JE & Vincent JFV (1982) The specific location of zinc in insect mandibles. *J. Exp. Biol.* **101**, 333–336 .

Hillerton JE & Vincent JFV (1983) Consideration of the importance of hydrophobic interactions in stabilising insect cuticle. *Int. J. Macromol.* **5**, 163–166 .

Hiltner A, Andersen JM & Baer E (1973) Dynamic mechanical analysis of poly-alpha-amino acids. Models for collagen. *J. Macromol. Sci. – Phys.* **8**, 431–443.

Hoeve CAJ & Flory PJ (1974) The elastic properties of elastin. *Biopolymers* **13**, 677–686.

Holmes OW (1907) *The Autocrat at the Breakfast Table*. In 'The People's Library' series. Cassell & Co: London.

Houghton PM & Sellen DB (1968) Dynamic mechanical properties of the cell walls of some green algae. *J. Exp. Bot.* **20**, 516–535.

Iizuka E (1966) Mechanism of fibre formation by the silkworm, *Bombyx mori* L. *Biorheology* **3**, 141–152.

Jackson AP (1986) The mechanical design of nacre. PhD Thesis, Reading University, UK.

Jackson AP, Vincent JFV, Briggs D, Crick RA, Davis SF, Hearn MJ & Turner RM (1987) The application of surface analytical techniques to the study of fracture surfaces in mother-of-pearl. *J. Mat. Sci. Lett.* **5**, 975–978.

Jackson AP, Vincent JFV & Turner RM (1988) The mechanical design of nacre. *Proc. R Soc.* B, **234**, 415–440.

Jackson AP, Vincent JFV & Turner RM (1989) A physical model of nacre. *Comp. Sci. Tech.* **36**, 255–266..

Jackson AP, Vincent JFV & Turner RM (1990) Comparison of Nacre with other ceramic composites. *J. Mat. Sci.* (in press).

Jennings JS & Macmillan NH (1984) A tough nut to crack. *J. Mat. Sci.* **21**, 1517–1524.

Jensen M & Weis-Fogh T (1962) Biology and physics of locust flight V. Strength and elasticity of locust cuticle. *Phil. Trans. R. Soc. Lond.* B **245**, 137–169.

Jeronimidis G (1976) The work of fracture of wood in relation to its structure. In *Wood Structure in Biological and Technological Research* (ed. P Baas, AJ Bolton & DM Catling). pp. 253–265. Leiden Botanical Series 3, Leiden University Press.

Jeronimidis G (1978) Fracture of wood and factors which influence it. *Proc. Oxford Congr. Int. Inst. Conservation of Historical & Artistic Work* (ed. NS Brommelle, A Moncriff & P Smith), pp. 7–10.

Jeronimidis G (1980) Wood, one of nature's challenging composites. In *The Mechanical Properties of Biological Materials* (*Symp. Soc. Exp. Biol.* 34) (ed. JFV Vincent & JD Currey), pp. 169–182. Cambridge University Press, Cambridge.

Jeronimidis G & Vincent JFV (1984) Composite materials. In *Connective Tissue Matrix* (ed. DWL Hukins). pp. 187–210. Macmillan, London.

Jin X, Jiang M, Yu T & Calvert P (1989) Preparation and properties of close-packed composite materials. *J. Mat. Sci.* **24**, 3416–3420.

Jones EI, McCance RA & Shackleton LRB (1935) The role of iron and silica in the structure of radular teeth of certain marine molluscs. *J. Exp. Biol.* **12**, 65–64.

Jones RW (1988) Sol-gel preparation of ceramics and glasses. *Met. Mat.* **4**, 748–751.

Kahler GA, Fisher FM & Sass RL (1976) The chemical composition and mechanical properties of the hinge ligament in bivalve molluscs. *Biol. Bull.* **151**, 161–181.

Kaleem K, Chertok F & Erhan S (1987) Novel materials from protein-polymer grafts. *Nature, Lond.* **325**, 328–329.

Kalischewski K & Schugerl K (1979) Investigation of protein foams obtained by bubbling. *Colloid Polymer Sci.* **257**, 1099–1110.

Kastelic J, Galeski A & Baer E (1978) The multicomposite structure of tendon. *Conn. Tiss. Res.* **6**, 11–23.

Katz EP & Li S-T (1973) Structure and function of bone collagen fibrils. *J. Mol. Biol.* **80**, 1–15.

Katz JL (1980) The structure and biomechanics of bone. In *The Mechanical Properties of Biological Materials* (*Symp. Soc. Exp. Biol.* 34) (ed. JFV Vincent & JD Currey), pp. 137–168. Cambridge University Press, Cambridge.

Katz JL (1988) Modelling the Young's moduli of dental composites. *J. Mat. Sci. Lett.* **7**, 133–134.

Kelly A (1973) *Strong Solids.* Clarendon Press, Oxford.

Kendall K & Fuller KNG (1987) J-shaped stress/strain curves and crack resistance of biological materials. *J. Phys. D: Appl. Phys.* **20**, 1596.

Ker RF (1977) Some structural and mechanical properties of locust and beetle cuticle. DPhil thesis, University of Oxford.

Ker RF (1981) Dynamic tensile properties of the plantaris tendon of sheep (*Ovis aries*). *J. Exp. Biol.* **93**, 283–302.

Ker RF, Alexander, RMcN & Bennet MB (1988) Why are mammalian tendons so thick? *J. Zool., Lond.* **216**, 309–324.

Khan AA (1988) Mechanical properties of fruits and vegetables, PhD Thesis, Reading University, UK.

Kinlock AJ (1987) *Adhesion and Adhesives, Science and Technology*. Chapman & Hall, London.

Kitchener AC (1987) Fracture toughness of horns and a reinterpretation of the horning behaviour of bovids. *J. Zool., Lond.* **213**, 621–639.

Kitchener AC (1988) An analysis of the forces of fighting of the blackbuck (*Antilope curvicapra*) and the bighorn sheep (*Ovis curvicapra*) and the mechanical design of the horns of bovids. *J. Zool., Lond.* **214**, 1–20.

Kitchener AC & Vincent JFV (1987) Composite theory and the effect of water on the stiffness of horn keratin. *J. Mat. Sci.* **22**, 1385–1389.

Klein JA, Hickey DS & Hukins DWL (1982) Computer graphics illustration of the operation of the intervertebral disc. *Eng. Med.* **11**, 11–15.

Knight DP (1968) Cellular basis for quinone tanning of the perisarc in the thecate hydroid *Campanularia (= Obelia) flexuosa* Hinks. *Nature, Lond.* **218**, 585–586.

Koehl MAR (1977) Mechanical diversity of connective tissue of the body wall of sea anemones. *J. Exp. Biol.* **69**, 107–125 .

Kramer EJ (1986) Crazing and cracking of polymers. In *Encyclopedia of Materials Science and Engineering* (ed. MB Bever), pp. 938–943. Pergamon, Oxford.

Lakes R (1987) Foam structures with a negative Poisson's ratio. *Science,* **235**, 1038–1040.

Lakes RS & Katz JL (1979) Viscoelastic properties and behaviour of cortical bone: Part II: Relaxation mechanisms. *J. Biomechanics* **12**, 679–687.

Lakes RS & Saha S (1979) Cement line motion in bone. *Science, NY* **204**, 501–503.

Lees C (1989) The milking machine and the cow's teat. Project Report, Reading University Dept. of Pure & Applied Zoology.

Levi C, Barton JL, Guillemet C, le Bras E & Lehuede P (1989) A remarkably strong natural glassy rod: the anchoring spicule of the *Monorhaphis* sponge. *J. Mat. Sci. Lett.* **8**, 337–339.

Lowenstam HA & Weiner S (1989) *On Biomineralisation*. The University Press, Oxford.

Lucas F & Rudall KM (1968) Extracellular fibrous proteins: the silks. In *Comparative Biochemistry* (ed. M Florkin & EH Stotz), 26B, pp. 475–558. Elsevier, Amsterdam.

Lugassy AA (1968) Mechanical and viscoelastic properties of bone and dentin in compression. PhD Dissertation, University of Pennsylvania.

Lusis J, Woodhams RT & Xanthos M (1973) The effect of flake aspect ratio on the flexural properties of mica-reinforced plastics. *Polym. Eng. Sci.* **13**, 139–145.

Maerkel K (1974) Morphologie der Seeigelzaehne V. Die Zaehne der Clypeastroidea (Echinodermata, Echinoidea). *Z. Morph. Tiere* **78**, 221–256.

Maerkel K, Kubanek F & Willgallis A (1971) Polykristalliner calcit bei Seeigeln (Echinodermata, Echinoidea). *Z. Zellforsch.* **119**, 355–377.

Maerkel K, Gorny P & Abraham K (1977) Microarchitecture of sea urchin teeth. *Fortschr. Zool.* **24**, 103–114.

Mai Y-W & Atkins AG (1989) Further comments on J-shaped stress–strain curves and the crack resistance of biological materials. *J. Phys. D: Appl. Phys.* **22**, 48–54.

Mann S (1983) Mineralization in biological systems. *Structure and Bonding* **54**, 125–174.

Mann S (1988) Molecular recognition in biomineralisation. *Nature, Lond.* **332**, 119–124.

Mann S & Calvert PD (1987) Biotechnological horizons in biomineralisation. *TIBTech* **5**, 309–314.

Mann S, Heywood BR, Rajam S & Birchall JD (1988) Controlled crystallisation of $CaCO_3$ under stearic acid monolayers. *Nature, Lond.* **334**, 692–695.

Mann S, Webb J & Williams RJP (1989) (eds) *Biomineralization – Chemical and Biochemical Perspectives.* VCH Publishers, Weinheim.

Manschot JFM & Brakkee AJM (1986). The measurement and modelling of the mechanical properties of human skin *in vivo* – II The model. *J. Biomechanics.* **19**, 517–521.

Mark RE (1967) *Cell Wall Mechanics of Tracheids.* Yale University Press, Yale.

Marquis PM (1989) Dental ceramics. *Met. Mat.* **5**, 145–148.

Marshall C & Gillespie JM (1977) The keratin proteins of wool, horn and hoof from sheep. *Austr. J. Biol. Sci.* **30**, 389–400.

Maroudas A (1975) Biophysical chemistry of cartilaginous tissues with specific reference to solute and fluid transport. *Biorheology* **12**, 233–248.

Mason B & Berry LG (1968) *Elements of Mineralogy.* Freeman, San Francisco.

Mattheck C (1989) *Engineering Components Grow like Trees*, pp. 1–75. Kernforschungszentrum Karlsruhe 4648.

Mattheck C, Huber-Betzer H & Keilen K (1989) *Die Anpassung eines Baumes an die Kontaktbelastung durch einen stein*, pp. 1–51. Kernforschungszentrum Karlsruhe 4562.

McCutchen CW (1975) Do mineral crystals stiffen bone by straitjacketing its collagen? *J. Theor. Biol.* **51**, 51–58.

Miller A (1980) A versatile body builder. *New Scientist* 14 February, 470–473.

Miller A (1984) Collagen: the organic matrix of bone. *Phil. Trans. R. Soc. Lond.* B **34**, 455–477.

Minke R & Blackwell J (1978) The structure of alpha-chitin. *J. Mol. Biol.* **120**, 167–181.

Mitchell JR & Blanshard JMV (1974) Viscoelastic behaviour of alginate gels. *Rheol. Acta.* **13**, 180–184.

Motokawa T (1983) Mechanical properties and structure of the spine-joint central ligament of the sea urchin, *Diadema setosum* (Echinodermata, Echinoidea). *J. Zool., Lond.* **201**, 223– .

Motokawa T (1988) Catch connective tissue: a key character for echinoderms' success. In *Echinoderm Biology* (ed. RD Burke, PV Mladenov, P Lambert & RL Parsley). Balkema, Rotterdam.

Mott NF (1964) Measuring 'hardness'. *New Scientist* no. **386**, 103–105.

Motta PJ (1987) A quantitative analysis of ferric iron in butterflyfish teeth (Chaetodontidae, Perciformes) and the relationship to feeding ecology. *Can. J. Zool.* **65**, 106–112.

Moyle DD & Gavens AJ (1986) Fracture properties of bovine tibial bone. *J. Biomechanics* **11**, 919–927.

Nilsson SB, Hertz CH & Falk S (1958) On the relation between turgor and tissue rigidity. II Theoretical calculations on model systems. *Physiol. Plant.* **11**, 818–837.

O'Donnell IJ & Inglis AS (1974) Amino acid sequence of a feather keratin from silver gull (*Larus novae-hollandiae*) and comparison with one from emu (*Dromaius novae-hollandiae*). *Austr. J. Biol. Sci.* **27**, 369–382.

O'Neill PL (1981) Polycrystalline echinoderm calcite and its fracture mechanics. *Science,* **213**, 646–648.

O'Neill PL (1989) Structure and mechanics of starfish body wall. *J. Exp. Biol.* **147**, 53–89.

Padawer GE & Beecher, N (1970) On the strength and stiffness of planar reinforced plastic resins. *Polym. Eng. Sci.* **10**, 185–192.

Papir YS, Hsu K-H & Wildnauer RH (1975) The mechanical properties of stratum corneum. The effect of water and ambient temperature on the tensile properties of newborn rat stratum corneum. *Biochim. Biophys. Acta* **399**, 170–180.

Payne AR (1965) In *Reinforcement of Elastomers* (ed. G Kraus). Wiley, New York.

Piez KA & Gross J (1959) The amino acid composition and morphology of some invertebrate and vertebrate collagens. *Biochim. Biophys. Acta* **34**, 24–39.

Piggott MR (1980) *Load-bearing Fibre Composites*. Pergamon, Oxford.

Pradas MM & Calleeja RD (1990) Reproduction in a polymer composite of some mechanical features of tendons and ligaments. In *High performance Biomaterials: a Guide to Medical/Pharmaceutical Applications* (ed. M Szycher). Technomic Press, USA.

Price HA (1981) Byssus thread strength in the mussel, *Mytilus edulis. J. Zool., Lond.* **194**, 245–.

Pryor MGM (1940) On the hardening of the ootheca of *Blatta orientalis. Proc. R. Soc. Lond.* B **128**, 378–398.

Pujol JP, Rolland M, Lasry S & Vinet S (1970) Comparative study of the amino acid composition of the byssus in some common bivalve molluscs. *Comp. Biochem. Physiol.* **34**, 193–201.

Purslow PP (1980) Toughness of extensible connective tissues. PhD thesis, University of Reading.

Purslow PP (1983) Positional variations in fracture toughness, stiffness and strength of descending thoracic pig aorta. *J. Biomechanics* **16**, 947–953.

Purslow PP (1987) The fracture behaviour of meat – a case study. In *Food Structure and Behaviour* (ed. JMV Blanshard & PJ Lillford), pp. 177–197. Academic Press, London.

Purslow PP (1989) Fracture of non-linear biological materials: some observations from practice relevant to recent theory. *J. Phys. D: Appl. Phys.* **22**, 854–856.

Purslow PP & Vincent JFV (1978) Mechanical properties of primary feathers from the pigeon. *J. Exp. Biol.* **72**, 251–260.

Purslow PP, Bigi A, Ripamonte A & Roveri N (1984) Collagen fibre reorientation around a crack in biaxially stretched aortic media. *Int. J. Biol. Macromol.* **6**, 21–25.

Ramsay JA (1971) *A Guide to Thermodynamics*. Chapman & Hall, London.

Rees DA (1977) *Polysaccharide Shapes*. Chapman & Hall, London.

Richards CW (1961) *Engineering Materials Science*. Chapman & Hall, London.

Richardson JS (1981) The anatomy and taxonomy of protein structure. *Adv. Prot. Chem.* **34**, 167–339.

Riley VR (1968) Fibre/fibre interaction. *J. Comp. Mater.* **2**, 436–446.

Roark RJ & Young WC (1975) *Formulas for Stress and Strain*, 5th edn. McGraw-Hill International, London.

Robertson B, Hillerton JE & Vincent JFV (1984) The presence of zinc or manganese as the predominant metal in the mandibles of adult stored product beetles. *J. Stored Prod. Res.* **20**, 133–137 .

Robertson DM & Smith DC (1978) Compressive strength of mandibular bone as a function of microstructure and strain rate. *J. Biomechanics* **11**, 455–471.

Rondell P (1970) Biophysical aspects of ovulation. *Biol. of Reprod. Suppl.* **2**, 64–89.

Rosen BW (1965) *Fibre Composite Materials*. pp. 37–75. Am. Soc. Metals.

de Rossi D (1989) Biomimetic approaches to the design of materials for artificial tactile perception. In *Proceedings of International Workshop on Intelligent Materials*, pp. 251–258. The Society of Non-Traditional Technology, Tokyo.

de Rossi D, Domenici C & Pastacaldi P (1986) Piezoelectric properties of dry human skin. *IEEE Trans.* E1-21, 511–517.

Roy DM (1987) New strong cement materials: chemically bonded ceramics. *Science, NY* **235**, 651–658.

Runham NW (1961) The histochemistry of the radula of *Patella vulgata. J. Micr. Sci.* **102**, 371–380.

Runham NW, Thornton PR, Shaw DA & Wayte RC (1969) The mineralisation and hardness of the radular teeth of the limpet, *Patella vulgata* L. *Z. Zellforsch.* **99**, 608–626.

Schofield R, Lefevre H & Shaffer M (1989) Complementary microanalysis of Zn, Mn and Fe in the chelicera of spiders and scorpions using scanning MeV-ion and electron microprobes. In *Nuclear Instruments and Methods in Physics Research* B40/41 pp. 698–701.

Schuerch H (1972) *Compressive Strength of Boron Metal Composites.* NASA Report CR202.

Scott J (1986) Molecules that keep you in shape. *New Scientist* 24 July, 49–53.

Serafini-Fracassini A, Field JM, Smith JW & Stephens WGS (1977) The ultrastructure and mechanics of elastic ligaments. *J. Ultrastruct. Res.* **58**, 244–251.

Shadwick RE & Gosline JM (1981) Elastic arteries in invertebrates: mechanics of the octopus aorta. *Science* **213**, 759–761.

Shamos MH & Lavine LS (1967) Piezoelectricity as a fundamental property of biological tissues. *Nature, Lond.* **211**, 267–269.

Shepherd W (1961) The susceptibility of hay species to mechanical damage. *Aust. J. Agric. Res.* **12**, 782–796.

Silyn-Roberts H & Sharp RM (1985a) Preferred orientation of calcite in the ratite and tinamou eggshells. *J. Zool., Lond.* **205**, 39–52.

Silyn-Roberts H & Sharp RM (1985b) Preferred orientation of calcite and aragonite in the reptilian eggshells. *Proc. R. Soc. Lond.* B **225**, 445–455.

Simkiss K & Wilbur KM (1977) The molluscan epidermis and its secretions. *Symp. Zool. Soc., Lond.* **39**, 35–76.

Simkiss K & Wilbur KM (1989) *Biomineralization: Cell Biology and Mineral Deposition.* Academic Press, San Diego.

Skertchly ARB (1964) Structural rheological phases in wool keratin. *J. Text. Inst.* 55 T154–T161.

Smeathers JE & Vincent JFV (1979) The mechanical properties of mussel byssus threads. *J. Moll. Stud.* **45**, 219–230.

Staines M, Robinson WH & Hood JAA (1981) Spherical indentation of tooth enamel. *J. Mat. Sci.* **16**, 2551–2556.

Stephens RC (1970) *Strength of Materials: Theory and Examples.* Edward Arnold, London.

Steudle E & Wieneke J (1985) Changes in water relations and elastic properties of apple fruit cells during growth and development. *J. Amer. Soc. Hort. Sci.* **110**, 824–829.

Strausberg RL & Link RP (1990) Protein-based medical adhesives. *TIBtech* **8**, 53–57.

Strout V, Lipke H & Geoghegan T (1976) Peptidochitodextrins of *Sarcophaga bullata*: molecular weight of chitin during preparation. In *The Insect Integument* (ed. HR Hepburn), pp. 43–61. Elsevier, Amsterdam.

Swamy RN (1979) Polymer reinforcement of cement systems Part I. Polymer impregnated concrete. *J. Mat. Sci.* **14**, 1521–1553.

Theron EP & Booysen PdeV (1966) Palatability in grasses. *Proc. Grassland Soc. S. Africa* **1**, 111–120.

Thompson, d'AW (1917) *On Growth and Form.* Cambridge University Press, Cambridge.

Timoshenko SP & Gere JM (1961) *Theory of Elastic Stability*, 2nd edn. McGraw-Hill, New York.

Timoshenko SP & Goodier JN (1970) *Theory of Elasticity*, 3rd edn. McGraw-Hill, New York.

Topping BHV (1989) Fully stressed design of natural and engineering structures. *Natuerliche Konstructionen (Proc. Sonderforschungsbereiches 230, Stuttgart)*, 2, 311–318.

Treloar, LRG (1975) *The Physics of Rubber Elasticity.* Oxford University Press, Oxford.

Tung MA, Staley LM & Richards JF (1968) Studies on the hardness of the hen's egg shell. *J. Agric. Eng. Res.* **13**, 12–18.

Urry DW (1983) What is elastin; what is not. *Ultrastruct. Pathol.* **4**, 227–251 .

Urban JPG & Maroudas A (1979) The measurement of fixed charge density in the intervertebral disc. *Biochim. Biophys. Acta* **586**, 166–178.

Vincent JFV (1975a) Locust oviposition: stress softening of the extensible intersegmental membranes. *Proc. R. Soc. Lond.* B **188**, 189–201.

Vincent JFV (1975b) How does the female locust dig her oviposition hole? *J. Ent.* A, **50**, 175–181.

Vincent JFV (1978) *Experiments with Biomaterials*. NCST, Trent Polytechnic, Nottingham .

Vincent JFV (1980a) The hardness of the tooth of *Patella vulgata* radula: a reappraisal. *J. Moll. Stud.* **46**, 129–133.

Vincent JFV (1980b) Insect cuticle: a paradigm for natural composites. In *The Mechanical Properties of Biological Materials* (*Symp. Soc. Exp. Biol.* 34) (ed. JFV Vincent & JD Currey), pp. 183–210, Cambridge University Press, Cambridge.

Vincent JFV (1982) The mechanical design of grass. *J. Mat. Sci.* **17**, 856–860.

Vincent JFV (1983) The influence of water content on the stiffness and fracture properties of grass leaves. *Grass Forage Sci.* **38**, 107–114.

Vincent JFV (1989) The relation between density and stiffness of apple flesh. *J. Sci. Food Agric.* **47**, 443–462.

Vincent JFV (1990) Fracture properties of plants. *Adv. Bot. Res.* **17**, 235–287.

Vincent JFV & Gravell K (1986) The mechanical design of kelp, *Laminaria digitata*. *J. Mat. Sci. Lett.* **3**, 310–312.

Vincent JFV & Hillerton JE (1979) The tanning of insect cuticle – a critical review and a revised mechanism. *J. Insect Physiol.* **25**, 653–658.

Vincent JFV & Jeronimidis G (1990) The mechanical design of fossil plants. In *Biomechanics in Evolution* (ed. JMV Rayner), (Soc. Exp Biol. Seminar Series). In press.

Vincent JFV & Owers P (1986) Mechanical design of hedgehog spines and porcupine quills. *J. Zool., Lond.* **210**, 55–75.

Vincent JFV & Wood SDE (1972) Mechanism of abdominal extension during oviposition in *Locusta. Nature, Lond.* **235**, 167–168.

Vollrath F & Edmonds DT (1989) Modulation of the mechanical properties of spider silk by coating with water. *Nature, Lond.* **340**, 305–307.

Wainwright SA (1988) *Axis and Circumference*. Harvard University Press, Cambridge, Mass.

Wainwright SA, Biggs WD, Currey JD & Gosline JM (1976). *Mechanical Design in Organisms*. Princeton University Press, Princeton.

Waite JH (1987) Nature's underwater adhesive specialist. *Int. J. Adhesion & Adhesives* **7**, 9–14.

Walker G, Yule AB & Ratcliffe J (1985) The adhesive organ of the blowfly, *Calliphora vomitoria*, a functional approach (Diptera, Calliphoridae). *J. Zool., Lond.* **205**, 297– .

Ward, IM (1971) *Mechanical Properties of Polymers*, Wiley, Chichester.

Waters NE (1980) Some mechanical and physical properties of teeth. In *The Mechanical Properties of Biological Materials* (*Symp. Soc. Exp. Biol.* 34) (ed. JFV Vincent & JD Currey), pp. 99–135. Cambridge University Press, Cambridge.

Watkins MR (1987) The development of a tough artificial composite based on antler bone. PhD Thesis, Reading University, UK.

Webb J, Macey DJ & Mann S (1989) Biomineralization of iron in molluscan teeth. In *Biomineralization, Chemical and Biochemical Perspectives* (ed. S Mann, J Webb & RJP Williams), VCH Publishers, New York.

Wegner G (1989) Advanced design criteria for organic and polymeric intelligent materials. In *Proceedings of International Workshop on Intelligent Materials*. The Society of Non-Traditional Technology, Tokyo, pp. 157–164.

Weiner S & Price PA (1986) Disaggregation of bone into crystals. *Calcif. Tissue Res.* **39**, 365–375.

Weiner S & Traub W (1984) Macromolecules in mollusc shells and their functions in biomineralization. *Phil. Trans. R. Soc. Lond.* B **304**, 425–434.

Weiner S & Traub W (1986) Organization of hydroxyapatite crystals within collagen fibrils. *FEBS Letters* **206**, 262–266.

Weis-Fogh T (1961a) Thermodynamic properties of resilin, a rubber-like protein. *J. Mol. Biol.* **3**, 520–531.

Weis-Fogh T (1961b) Molecular interpretation of the elasticity of resilin, a rubber-like protein. *J. Mol. Biol.* **3**, 648–667.

Weis-Fogh T & Amos WB (1972) Evidence for a new mechanism of cell motility. *Nature, Lond.* **236**, 301–304.

Weis-Fogh T & Andersen SO (1970) New molecular model for the long-range elasticity of elastin. *Nature, Lond.* **227**, 718–721.

White SW, Hulmes DJS, Miller A & Timmins PA (1979) Collagen-mineral axial relationship in calcified turkey leg tendon by x-ray and neutron diffraction. *Nature, Lond.* **266**, 421–425.

Whitehouse WJ & Dyson ED (1974) Scanning electron microscope studies of trabecular bone in the proximal end of the human femur. *J. Anat.* **118**, 417–444.

Whorlow RW (1980) *Rheological Techniques*. Ellis Horwood, Chichester.

Wilkie IC (1978) Nervously mediated change in the mechanical properties of a brittlestar ligament. *Marine. Behav. Physiol.* **5**, 289–306.

Willis A (1989) Design and development of an instrumented microtome. PhD Thesis, University of Reading.

Willson MC & Dower RJ (1988) Ceramic matrix composites. *Met. Mat.* **4**, 752–756.

Woodhead-Galloway J (1975) Collagen – the universal body builder. *New Scientist* 11 September, 582–584.

Woodhead-Galloway J (1981) The body as an engineer. *New Scientist* 10 June 772–775.

Wortmann F-J (1987) The viscoelastic properties of wool and the influence of some specific plasticisers. *Coll. Polym. Sci.* **265**, 126–133.

Wu H, Spence RD & Sharpe PJH (1988) Plant cell wall elasticity II: Polymer elastic properties of the microfibrils. *J. Theor. Biol.* **133**, 239–253.

Yoon HS (1987) Synthesis of fibres by growth-packing. *Nature, Lond.* **326**, 580–582.

Young GA & Crisp DJ (1982) Marine animals and adhesion. In *Adhesion – 6* (ed. KW Allen), pp. 279–313. Applied Science Publishers, Barking.

Zweben C, Smith WS & Wardle MW (1979) Test methods for fiber tensile strength, composite flexural modulus and properties of fabric-reinforced laminates. In *Composite Materials: Testing and Design* (5th conference) (ed. SW Tsai), pp. 244–262. ASTM special publication 674.

Subject Index